高等学校酿酒工程专业教材

酿酒工程专业英语

聂 聪 张浩军 刘 君 主编

中国轻工业出版社

图书在版编目（CIP）数据

酿酒工程专业英语/聂聪，张浩军，刘君主编. —北京：中国轻工业出版社，2023.5
ISBN 978-7-5184-4265-2

Ⅰ.①酿… Ⅱ.①聂… ②张… ③刘… Ⅲ.①酿酒—英语—高等学校—教材 Ⅳ.①TS261.4

中国国家版本馆CIP数据核字（2023）第023823号

责任编辑：江 娟 李 蕊
策划编辑：江 娟　　责任终审：许春英　　封面设计：锋尚设计
版式设计：砚祥志远　责任校对：吴大朋　　责任监印：张 可

出版发行：中国轻工业出版社（北京东长安街6号，邮编：100740）
印　　刷：三河市国英印务有限公司
经　　销：各地新华书店
版　　次：2023年5月第1版第1次印刷
开　　本：787×1092　1/16　印张：11.25
字　　数：326千字
书　　号：ISBN 978-7-5184-4265-2　定价：38.00元
邮购电话：010-65241695
发行电话：010-85119835　传真：85113293
网　　址：http://www.chlip.com.cn
Email：club@chlip.com.cn
如发现图书残缺请与我社邮购联系调换
200095J1X101ZBW

全国高等学校酿酒工程专业教材
编委会

主　任　　徐　岩（江南大学）
副主任　　李　崎（江南大学）
　　　　　石贵阳（江南大学）
　　　　　李　华（西北农林科技大学）
　　　　　张文学（四川大学锦江学院）
　　　　　肖冬光（天津科技大学）
　　　　　段长青（中国农业大学）
　　　　　宋全厚（中国食品发酵工业研究院）
顾　问　　王延才（中国酒业协会）
　　　　　宋书玉（中国酒业协会）
　　　　　金征宇（江南大学）
　　　　　顾国贤（江南大学）
　　　　　章克昌（江南大学）
　　　　　赵光鳌（江南大学）
　　　　　夏文水（江南大学）
委　员　　（按姓氏拼音排序）
　　　　　白艳红（郑州轻工业大学）
　　　　　陈忠军（内蒙古农业大学）
　　　　　杜金华（山东农业大学）
　　　　　范文来（江南大学）
　　　　　付桂明（南昌大学）
　　　　　龚国利（陕西科技大学）
　　　　　何　惠（茅台学院）

黄名正（贵州理工学院）
寒华丽（华南农业大学）
李宪臻（大连工业大学）
李　艳（河北科技大学）
廖永红（北京工商大学）
刘世松（滨州医学院）
刘新利（齐鲁工业大学）
罗惠波（四川轻化工大学）
毛　健（江南大学）
邱树毅（贵州大学）
单春会（石河子大学）
孙厚权（湖北工业大学）
孙西玉（河南牧业经济学院）
王　栋（江南大学）
王　君（山西农业大学）
文连奎（吉林农业大学）
贠建民（甘肃农业大学）
赵金松（四川轻化工大学）
张　超（宜宾学院）
张军翔（宁夏大学）
张惟广（西南大学）
周裔彬（安徽农业大学）
朱明军（华南理工大学）

前　言

《酿酒工程专业英语》是在中国轻工业出版社的组织协调下，结合中国酿酒学科高校联盟多所大学的酿酒专业英语教学经验编写而成，在内容的安排上力求做到科学性与系统性的统一，内容涉及酿造酒中的啤酒、葡萄酒、黄酒，蒸馏酒中的白兰地、威士忌、伏特加和中国的白酒等，以及配制酒中的果酒（配制型）、蛋奶酒、咖啡酒的原料、发酵及蒸馏等酿造工艺及设备的相关内容。

本教材的编写目标是培养学生理解并较系统地掌握所有与酒类酿造工艺及设备相关的专业词汇，特别是核心的三大酒类——啤酒、葡萄酒和蒸馏酒，提高专业英语文献的阅读和理解能力。各章节均选自国内外原版资料，并有单词、注释、阅读材料等。在编排上以酒的种类为主线，结合其酿造工艺、酿酒文化、酿造设备等内容，尽量反映出各种酒的历史、发展及特点等，具有广泛的适用性。

本教材由齐鲁工业大学（山东省科学院）聂聪、张浩军，四川轻化工大学刘君担任主编，齐鲁工业大学教材建设基金资助。由刘君负责中国白酒部分的编写工作，其余部分由聂聪、张浩军负责。

本教材适用于普通高等院校酿酒工程、葡萄酒工程、食品科学与工程等专业的大学生作为教材使用，也可供从事酒类酿造、食品加工以及相应的管理人员作为参考。

由于编者水平和各方面条件限制，教材中难免存在不足之处，恳请读者提出宝贵意见。

编者
2022 年 12 月

目 录

Chapter 1 Brewed Alcoholic Beverages

Unit 1 Beer ·· 002
   Lesson 1　Barley and Malt ·· 002
   Lesson 2　Hops ··· 005
   Lesson 3　Yeast Management ·· 009
   Lesson 4　Wort Preparation ·· 012
   Lesson 5　Fermentation and Maturation ··· 017
   Lesson 6　Beer Flavor and Stability ··· 020

Unit 2 Wine ·· 023
   Lesson 1　The Variety of Grapes ·· 023
   Lesson 2　Viticulture ·· 026
   Lesson 3　Red Wine Vinification ·· 029
   Lesson 4　White Wine Vinification ··· 034
   Lesson 5　Wine Chemistry ··· 038
   Lesson 6　Wine Sensory Properties ·· 040
   Lesson 7　Sparkling Wine ··· 043

Unit 3 Rice Wine ··· 046
   Lesson 1　Sticky Rice ··· 046
   Lesson 2　Jiuqu ·· 048
   Lesson 3　Shaoxing Rice Wine ··· 051
   Lesson 4　Jimo Lao Jiu Rice Wine ··· 053
   Lesson 5　Sake ·· 056

Chapter 2 Distilled Alcoholic Beverages

Unit 1 Chinese Baijiu ·· 060
   Lesson 1　Introduction ··· 060
   Lesson 2　Daqu ·· 066

Lesson 3	Tasting	080
Lesson 4	Packaging	084
Lesson 5	Distillation and Maturation	089
Lesson 6	Blending Technique	092

Unit 2 Brandy · 095
Lesson 1	Introduction	095
Lesson 2	Distilled Technique	097
Lesson 3	Barrel and Aging	099
Lesson 4	Blending Technique	102

Unit 3 Whisky · 105
Lesson 1	Introduction	105
Lesson 2	Distilled Technique	108
Lesson 3	Barrel and Aging	111
Lesson 4	Blending Technique	113

Unit 4 Other Distilled Alcoholic Beverages · 117
Lesson 1	Vodka	117
Lesson 2	Rum	120
Lesson 3	Tequila	122
Lesson 4	Gin	125

Chapter 3 Blended Alcoholic Beverages

Unit 1 Liqueur · 130
Lesson 1	Fruit Liqueur	130
Lesson 2	Eggnog Liqueur	133
Lesson 3	Coffee Liqueur	137
Lesson 4	Anisette	140

Unit 2 Other Blended Alcoholic Beverages · 144
Lesson 1	Vermouth	144
Lesson 2	Martini	148

Vocabulary · 151

References · 165

Chapter 1
Brewed Alcoholic Beverages

Unit 1 Beer
Unit 2 Wine
Unit 3 Rice Wine

Unit 1　Beer

Lesson 1　Barley and Malt

Objectives

At the end of this lesson, students should be able to:
1. Introduce briefly what Barley (*Hordeum vulgare* L.) is.
2. Know where barley originated.
3. Know what the function of barley is.

Text

In terms of production, barley (*Hordeum vulgare*) is the fourth most important crop in the world. Barley (*Hordeum vulgare* L.) is one of the Neolithic founder crops. It is a flowering plant belonging to the family Poaceae or Gramineae (herbs) that is cultivated in temperate zone at 350~4050m above sea level, and evolved from the wild barley (*H. spontaneum*). Species of barley consist of diploid ($2n = 2x = 14$), tetraploid ($2n = 4x = 28$), and hexaploid ($2n = 6x = 42$) cytotypes. Barley constitutes the fourth most important grain crop in the world after wheat, rice and maize. Barley grain is used as livestock feed and forage, malt beverages, human food, soil improvement and has medicinal value, but barely is considered as a highly-needed crop of the present era. Common barley hails originally from Western Asia and North Africa. It is one of the earliest documented agricultural grains, dating back to the Neolithic period (8500 years ago) in the Nile Delta portion of the Fertile Crescent. Barley is a rich source of proteins, B vitamins, niacin, minerals and fiber dietary; also, it is a good source of manganese and phosphorus. Raw barley consists of carbohydrates (78%), proteins (10%), water (10%) and fat (1%). It is the main raw material for making natural starch, starch derivatives, high fructose syrup, etc. In addition, barley is also the best raw material for beer and whisky. In a word, barley is rich in nutrition and has health effects. It is an ideal raw material for processing popular, economical, convenient and therapeutic food.

Malt is a kind of high amylase additive. Adding malt into bread flour with low amylase activity can improve baking properties, and can also be used as flavor additive to make various foods. Malt can be used to produce high protein malt powder (making high nutritional food) and low protein malt powder (brewing excellent beer). It can also produce saccharified or non saccharified malt concentrate (for candy and preserves, as a suitable carrier for pharmaceutical), cereal syrup and other products, and can be used for baking food, breakfast food, baby food and rehabilitation food. It can also produce alcohol, beer and malt vinegar and other addictive food.

The most common and useful way for ensuring that a malting plant has been properly assessed and is in control of food safety is to have a Hazard Analysis and Critical Control Points (HACCP) system in place.

Table 1-1   Examples of prerequisite control measures that form a basis for a HACCP protocol

Area for control	Methods of control
Grain intake and storage	Strict control of reception and storage of grain according to best practice guidelines, e.g. moisture. Strict terms and conditions for supply and supplier evaluation procedures. Use of an assured grain supply scheme. Due diligence sampling to assure controls effective
Housekeeping	Must be of a high standard throughout the maltings. For example, different coloured brooms should be used for areas with different risks, e.g. green for outside roadways and paths, blue for inside the plant and red for germination boxes, thus keeping cleaning of food and pedestrian areas separated
Jewellery	No loose jewellery, watches, spectacles or anything else that could fall into the product
Glass	Generally minimal glass allowed on site. All glass or brittle plastics must be on an audited register and risk assessed
Smoking	If allowed on site it must be restricted to a defined area away from production
Personal hygiene	Hand washing or sanitizing stations must be provided at entrances to food areas, e.g. germination boxes
Cleaning	Chemicals must be food grade. Cleaning schedules must be documented
Pest control	Must be extremely thorough and any actions identified must be rectified immediately
Visitors and contractors	Brief induction in food safety at reception to site
Personnel hygiene	Basic hygiene rules are mandatory and staff are restricted from work if they or their close family members are ill

This table lists some of the key prerequisites that must be in place for a suitable HACCP system for malting (Table 1-1). It is not intended to be exhaustive, but demonstrates the foundation on which a HACCP study can build.

Questions on Text

Briefly introduce what malt is made of.

Words to Watch

crop [krɒp] n. 农作物；庄稼；产量 vt. 种植；收割；修剪；vi. 收获
poaceae/gramineae n. 禾本科
diploid [ˈdɪplɔɪd] n. 二倍体
tetraploid [ˈtetrəˌplɔɪd] n. 四倍体
hexaploid [ˈheksəplɔɪd] n. 六倍体
production [prəˈdʌkʃən] n. 生产，制作；产量
agriculture [ˈæɡrɪkʌltʃə] n. 农业
cultivate [ˈkʌltɪveɪt] v. 培养，培育；种植
protein [ˈprəʊtiːn] n. 蛋白质
carbohydrate [ˌkɑːbəʊˈhaɪdreɪt] n. 碳水化合物；糖类；含碳水化合物的食物
vitamin [ˈvɪtəmɪn] n. 维生素
niacin [ˈnaɪəsɪn] n. 烟酸
mineral [ˈmɪnərəl] n. 矿物质
fiber dietary [faɪbə ˈdaɪətəri] 膳食纤维
manganese [mæŋɡəˈniːz] n. 锰
phosphorus [ˈfɒsfərəs] n. 磷

Phrases and Patterns

in terms of 依据
evolve from 从……进化而来
be used as 被用作……

Supplementary reading

Germination represents a key developmental phase in plant physiology, which is accompanied by many significant changes in metabolite contents, so it plays a central role in the beer brewing industry.

Moreover, endogenous hydrolases in barley grains are also synthesized and/or activated at this stage for hydrolyzing macromolecular substances stored in the barley endosperm. Therefore, stimulating both barley germination and hydrolase activity in malt is important for increasing wort yield and the quality of the beer product. Several additives have been used in the malting process to obtain excellent germination ability and wort yields. However, most of these additives increase production costs, food safety problems and environmental pollution.

Words to Watch

endogenous [enˈdɒdʒənəs] adj. 内生的
hydrolase [ˈhaɪdrəleɪs] n. 水解酶
synthesize [ˈsɪnθəsaɪz] vt. 综合，合成
hydrolyze [ˈhaɪdrəlaɪz] v. 水解
macromolecular [ˌmaːkrəʊˈmɒlɪˌkjuːlə] adj. 大分子的
endosperm [ˈendəʊspɜːm] n. 胚乳
germination [ˌdʒɜːmɪˈneɪʃən] n. 发芽，萌发

Lesson 2　Hops

Objectives

At the end of this lesson, students should be able to:
1. Briefly introduce what hops are.
2. Know where hops originated.
3. Know which countries are the major producers of hops in the world.

Text

The hops are dioecious, perennial, climbing vines. This genus belongs to the Cannabaceae family of the Urticales order which in 2003 was incorporated into the natural order of Rosales. For years it was believed that the genus *Humulus* was only represented by two species, the 'common hop', *Humulus lupulus* L., and the 'Japanese hop', *H. japonicus* Sieb. et Zucc. In 1936, the species *Humulus yunnanensis* Hu was first described; however, it remained a relatively unknown species which is thought to have originated at high elevations in the Yunnan province of southern China. Previous to Small's study, it had been almost universally unappreciated and identified as *H. lupulus*. It was in 1978 that Small identified *H. yunnanensis* as a species of its own. Currently more is known of the *H. japonicus* species; this is an annual plant indigenous in China, Japan and the neighbouring islands. *H. japonicus* is widely cultivated as a strong climber, often used in gardens as a decorative, leafy screen.

The hop cones of the female plant of the common hop species *Humulus lupulus* L. are grown almost exclusively for the brewing industry. About 97% of worldwide cultivated hops are destined for brewing purposes. The world hop production is dominated by Germany and the USA. The hop production of both countries accounts for about 75%~80% of the total hop output. Successful cultivation of the hop plant requires optimal growth conditions, especially with respect to the length of daylight, summer temperature, annual rainfall and soil fertility. The hop plant is grown throughout most moderate climate regions of the world; these are located between latitudes 35° and 55° of the Northern and Southern Hemispheres. More than 60% of the hop area under cultivation is located in Germany and the USA. The largest hop growing areas include the Hallertau region in Germany and the states of Washington, Oregon and Idaho in the USA. Other hop growing countries are the Czech Republic, Poland, Slovenia, England, Ukraine, China, South Africa, Australia and New Zealand.

Hop Constituents

Hops contain hundreds of components, but of particular interest are resins, oils, and polyphenols.

Hop Resins

Hop resins are subdivided into hard and soft, based on their solubility. Hard resins are of little significance, as they contribute nothing to the brewing value, while soft resins contribute to the flavoring and preservative properties of beer. Alpha and beta acids are two compounds present in the soft resins and are responsible for bitterness. Alpha acids are responsible for about 90% of the bitterness in beer. Magnesium, carbonate, and chloride ions can also accentuate hop bitterness.

Alpha Acids: Alpha acids are the precursors of beer bitterness since they are converted into iso-alpha acids in the brew kettle. The three major components of alpha acids are humulone, cohumulone, and adhumulone.

Beta Acids: Hops also contain a second group of acids known as the beta acids. The beta acids (lupulone, colupulone, and adlupulone) are only marginally bitter.

Hop Oils

Although hops that have high alpha acid content are preferred for their bittering and flavoring properties, hops are also selected for the character of their oils. Oils are largely responsible for the characteristic aroma of hops and, either directly or indirectly, for the overall perception of hop flavors. Hops selected for character of their oil content are often referred to as aroma or "noble" type hops. Oils also tend to make a beer's bitterness a little more pronounced and to enhance the body or mouth feel of the beer.

Hop Polyphenols

Polyphenols found in hops include the anthocyanogens, tannins, and catechins. Some polyphenols act as antioxidants, protecting beer against oxidation, while others contribute to beer color and haze formation. Polyphenols may also cause an unpleasant astringency. Significant proportions are removed during boiling by precipitation with proteins.

Besides whole hops (hop cones), other hop products are used. The international hop industry currently recognizes four general types of hop products. They are non-isomerized hop products, isomerized hop products, and hop oil products. The ones most commonly traded in the commercial market are pellets and extracts.

Questions on Text

What conditions do hops need to grow?
Which hop product is the most commonly used right now?
What constituents do hops contain?

Words to Watch

dioecious [daɪˈiːʃəs] adj. 雌雄异体的
perennial [pəˈrɛnɪəl] n. 多年生植物　adj. 长期的，持久的
represent [ˌrɛprɪˈzɛnt] vt. 代表，表现，描绘
relatively [ˈrɛlətɪvli] adv. 相对地，相当地
optimal [ˈɒptɪməl] adj. 最理想的；最佳的
grown [ɡrəʊn] adj. 长大的；成年的，成熟的　v. 生长（grow 的过去分词）
hop resins [hɒp ˈrɛzɪnz] 酒花树脂
Humulus lupulus 啤酒花（大麻科葎草属）
humulone [hjuːmjʊˈlən] n. 葎草酮
cohumulone [kəʊhjuːmjʊˈlən] n. 合葎草酮
adhumulone [ədhjuːmjʊˈlən] n. 加葎草酮
lupulone [ˈluːpjʊləʊn] n. 蛇麻酮；蛇床酮

colupulone [kəluːpjʊlˈwʌn] n. 类蛇麻酮
adlupulone [ædˈluːpjuləun] n. 聚蛇麻酮；伴蛇麻酮
polyphenol [pɒlɪˈfiːnɒl] n. 多酚；多酚类
anthocyanogen [ˈænθəsaɪənədʒɪn] n. 花青素类
tannin [ˈtænɪn] n. 单宁酸；鞣酸
catechin [ˈkætɪtʃɪn] n. 儿茶酸；焦儿茶酸
hop cone 蛇麻球果；啤酒花果穗

Phrases and Patterns

be made up of 由……组成
be dominated by 受……支配
be located between 位于

Supplementary reading

In most commercial hop growing areas worldwide the seed content in hops is regulated. Male hops are physically removed from the hop fields to avoid the fertilization of the female plants and, thus, the production of seeds. In most parts of the world, seeds are considered by brewers to be undesirable. It is believed that oxidation of the seed fatty acids produce off-flavours in beer. Further, it has been proven that seedless hops are generally richer in essential oils and resins than seeded ones (i.e. higher brewing value). However, male plants are essential in hop breeding programmes to develop new varieties through controlled hybridization.

Hop utilization describes the efficiency of alpha acid utilization as iso-alpha acids in the finished beer. Not all of the bitterness potential from the alpha acid in the hop is utilized, given the losses that occur throughout the brewing process. Therefore, the utilization of the bitter substances rarely exceeds 40% in commercial breweries and is often as low as 25%. The main reason for the low utilization has to do with low pH of the wort (i.e., pH of 5.0~5.5), with more efficient utilization occurring at a higher pH (i.e., pH of 10.0~11.0) (Table 1-2).

Table 1-2　　　　　　　　　　　Hops composition

Principle Components	Concentration (%w/w)
Cellulose and lignin	40.0 ~ 50.0
Protein	15.0
Alpha acid	2.0 ~ 17.0
Beta acid	2.0 ~ 10.0

续表

Principle Components	Concentration (%w/w)
Water	8.0 ~ 12.0
Mineral	8.0
Polyphenol and tannin	3.0 ~ 6.0
Lipid and fatty acid	1.0 ~ 5.0
Hop oil	0.5 ~ 3.0
Monosaccharide	2.0
Pectin	2.0

* European Brewery Convention. Hops and Hop Products, Manual of Good Practice. Nurnberg: Fachverlag Hans Carl, 1997.

Lesson 3 Yeast Management

Objectives

At the end of this lesson, students should be able to:

1. Know what the types of beer are.

2. Know what substances non-*saccharomyces* yeast will produce in the self-fermentation process.

3. Know why non-*saccharomyces* yeast fermentation will turn more alcohol fermentation intermediates to secondary metabolites.

Text

Yeast are single-celled microorganisms that reproduce by budding. They are biologically classified as fungi and are responsible for converting fermentable sugars into alcohol and other byproducts. There are literally hundreds of varieties and strains of yeast. In the past, there were two types of beer yeast: ale yeast (the "top-fermenting" type, *Saccharomyces cerevisiae*) and lager yeast (the "bottom-fermenting" type, *Saccharomyces uvarum*, formerly known as *Saccharomyces carlsbergensis*). Today, as a result of recent reclassification of *Saccharomyces* species, both ale and lager yeast strains are considered to be members of *S. cerevisiae*. However, throughout this chapter and the book, both ale and lager yeasts will be referred to those strains previously classified as members of *S. cerevisiae* and *S. carlsbergensis*, respectively.

Top-fermenting yeasts are used for brewing ales, porters, stouts, Altbier, Kölsch, and wheat beers. Some of the lager styles made from bottom-fermenting yeasts are Pilsners, Dortmunders, Märzen, Bocks, and American malt liquors.

Beers are broadly classified as either ale or lager depending on the yeast used (*Saccharomyces cerevisiae* or *S. pastorianus*, respectively) and fermentation conditions. Nevertheless, modern brewers today embrace the use of novel ingredients including spices, herbs and fruits to alter a beer's flavour, creating what is regarded as craft or specialty beers. However, when it comes to a beer's characteristic flavour, the yeast used remains to be the most significant contributor. This is attributed to the numerous flavour-active substances generated by yeast.

Due to the abundance and dominance of yeast strains from the *Saccharomyces* genus during spontaneous fermentations, they have been intuitively selected and studied over generations. In contrast, non-*Saccharomyces* yeasts have often been disregarded for their overproduction of off flavour compounds such as acetic acid, diacetyl and 2, 3-butanediol. Despite this, the rising demand of new specialty beers has driven researchers to isolate and re-evaluate the potential of non-*Saccharomyces* yeasts for beer fermentation.

Compared to *Saccharomyces* species, non-*Saccharomyces* yeast fermentations divert more intermediates of alcoholic fermentation towards secondary metabolite generation (e.g. esters, fusel alcohols, organic acids), resulting in correspondingly less biomass synthesis and ethanol production. Consequently, this results in an abundance of species and/or strain-specific fermentative by-products which strongly characterise and differentiate yeast species/ strains.

There are several important concepts about brewing yeast. The term "flocculation" refers to the tendency to form clumps of yeast called flocs. The flocculation characteristics of yeast are of great importance. The flocs (yeast cells) descend to the bottom in the case of bottom-fermenting yeasts or rise with carbon dioxide bubbles to the surface in the case of top-fermenting yeasts.

Attenuation refers to the percentage of sugars converted to alcohol and carbon dioxide, as measured by specific gravity. Most yeasts ferment the sugars glucose, sucrose, maltose, and fructose. To achieve efficient conversion of sugars to ethanol (good attenuation) requires the yeast to be capable of completely utilizing the maltose and maltotriose.

Yeast mutations are a common occurrence in breweries, but their presence may never be detected. Usually the mutant has no adverse effect since it cannot compete with normal yeast and generally disappears rapidly.

Yeast degeneration refers to the gradual deterioration in performance of the brewing yeast. Yeast degeneration has a harmful effect on the course of brewing fermentations. It is characterized by some of the following symptoms: sluggish fermentations, premature cessation of fermentation (resulting in high residual fermentable levels in beer), gradual lengthening of fermentation times, and poor foam or yeast head formation.

Questions on Text

What results in abundant species and/or fermentation by-products of specific strains?

Words to Watch

ale [eɪl] n. 爱尔啤酒，艾尔啤酒
lager [ˈlɑːgə] n. 拉格啤酒，贮藏啤酒
Saccharomyces cerevisiae 酿酒酵母
ingredient [ɪnˈgriːdiənt] n. （混合物的）组成部分；配料
specialty beers 特色啤酒
herb [hɜːb] n. 香草，药草
flavour [ˈfleɪvə] n. 香味；滋味 vt. 给……调味
diacetyl [daɪəˈsiːtɪl] n. 双乙酰
acetic acid 乙酸，醋酸
secondary metabolite 次级代谢产物
biomass [ˈbaɪəumæs] n. （单位面积或体积内的）生物量
strain [streɪn] n. 菌株
domestication [dəʊˌmestɪˈkeɪʃən] n. 驯化
flocculation [flɒkjʊˈleɪʃən] n. 絮凝，絮结产物
attenuation [əˌtenjuˈeɪʃən] n. 发酵度
relative density 相对密度

Phrases and Patterns

result in 导致
an abundance of 大量的
be attributed to 归因于

Supplementary reading

Text

Here, we describe, in natural ploidy, high-quality sequencing, de novo assembly, annotation and extensive phenotypic analysis of the 157 *S. cerevisiae* strains used for industrial production of beer, wine, bread, spirits, sake and bioethanol. Our data indicate that industrial yeasts are genetically and phenotypically different from wild strains and are only derived from limited ancestral strains that have been adapted to an artificial environment. They are further divided into five clades: one containing Asian strains such as sake yeast, one mainly containing wine yeast, one mixed clade containing bread and other yeast, and two beer yeast families. Although most clades lack a strong geographic substructure, one of the beer clades contains geographically separated subpopulations of strains of beer in the European continent (Belgium / Germany), the United Kingdom, and the recently settled American brewer's yeast sub-line. Interestingly, these beer yeast lineages have obvious and profound domestication characteristics than other lineages. The change from the variable, complex and often harsh environment encountered in nature to a more stable, nutrient-rich beer culture medium is conducive to the special adaptation of brewer's yeast, but also leads to genome decay, aneuploidy and function loss of sexual cycle. Specifically, we found evidence of active human selection, mainly through mutation and repetition of the MAL (maltose) gene and meaningless mutations in PAD1 and FDC1, as evidenced by the fusion evolution of efficient fermentation of beer specific carbon sources. When producing 4-vinylguaiacol (4-VG), an unpleasant odor in beer. Our results further show that the domestication of beer yeast began hundreds of years ago, and the production of beer was first reported very early, but before the discovery of microorganisms. In summary, our results reveal that today's industrial yeast is the product of human domestication for hundreds of years, and provides new resources for further selection and breeding of superior varieties.

Lesson 4 Wort Preparation

Objectives

At the end of this lesson, students should be able to:
1. Know how wort is made.
2. Know how to clarify wort.
3. Know what the function of yeast is in the fermentation process.

Text

Wort was produced from malt, starch, rice, corn grits, and hop pellet. Wort clarification was carried out as follows; perlite was added to wort at a dosage of 100mg/L and the wort was then filtered with filter sheets.

It is well known that sulfite is excreted from yeasts during fermentation and remains in the finished beer. Many researchers have studied factors which influence sulfite formation by yeast during fermentation for the control of sulfite level in beer and for the stabilization of beer. These factors have been reported as being the yeast strain, wort pH, wort aeration, wort trub content, wort composition and nutritional deficiency, and fermentation temperature. However, there are very few studies to confirm the relationship between wort fermentation and flavor stability of the resulting beer. Therefore, it is still a common question for brewers whether sulfite naturally produced by yeast during wort fermentation can contribute to flavor stability of the finished beer.

Before starting the brewing process, barley is artificially induced to germinate and dried. This process, known as malting, allows the maturation of enzymes that digest complex starches in the grain into simple fermentable sugars. These sugars will be used later during mashing. Although malted barley is the most important cereal, wheat, wheat malt, corn, rice and millet are often used. The transition of malted barley to wort begins by grinding the grains. The object of milling is to split the husk, preferably lengthwise, in order to expose the starchy endosperm for milling and allow for efficient extraction and subsequent filtration of the wort. However, it is necessary to compromise between the requirements for extraction and filtration. Although a fine grind potentially yields more extract, it can lead to subsequent filtration problems and a loss of extract in the spent grains. Malt milling is usually done by either dry or wet milling.

The selection of the type of milling and employment of a brewery-specific milling process is determined by the size distribution of malt kernels, their modification, moisture content, the mashing methods, and the wort separation method.

Malt Dry Milling

The most commonly used mills in breweries are dry grist mills. Mills are usually either of the roller type or based on impact, i.e. hammer mills. If the wort separation process is a mash/lauter tun, roller mills are employed. Hammer mills are largely used for the later generation of mash filters and continuous brewing systems. However, new generation mash filters, for example, Meura 2001, require the use of hammer mills.

Malt Wet Milling

In a wet milling operation, the whole uncrushed malt is pre-steeped in hot water to the point where the husks reach a water content of approximately 20% and the endosperm remains nearly dry, which results in a semi-plastic, almost pasty consistency. The duration and temperature of steeping depends on the modification and the moisture content of the malt. The steeped malt is then passed through a two-roll mill in which the endosperm is squeezed out of the surrounding husk, leaving it fully available to the subsequent actions of mashing. The husk remains tough and intact for its eventual role as the filter medium for wort separation. At this stage the grist is diluted, filtered and brought to boiling for one to two hours. In this process, hops or hop-derived products, such as hop pellets or hop essential oils are added. Hops possess a characteristic flavour and aroma that exert a strong influence on the final product and so its dosage needs to be adjusted to the profile of the desired beer and integrated harmonically in the matrix during the beer maturation. At the end of the boil, the resulting wort contains coagulated proteins or 'trub' and suspended fragments of hops that must be removed, often using a 'whirlpool tank' or 'settling tank'. After cooled and aerated, the clear wort is then pumped to the fermentation tanks, where yeasts are added. During fermentation, yeasts take up amino acids and sugars from the wort. The sugars are metabolized, under anaerobic conditions, and converted mainly in ethanol and carbon dioxide. Additionally, this carbohydrates fermentation generates a typical fingerprint of volatile metabolites at relatively low levels, like aldehydes, ketones, higher alcohols, organic acids, and esters, which are called 'fermentation by-products' or 'congeners'. The temperature and time of fermentation are the key factors in the quality of the beer. Fermentation takes place at $7 \sim 14\,°C$, for lager beers, or $16 \sim 18\,°C$ for ale beers. In the particular case of lambic type, the wort is contaminated by yeasts present in the atmosphere and fermentation occurs spontaneously during $1 \sim 2$ years at room temperature.

Mashing

Mashing is the process by which sweet wort is prepared. It involves 'mashing in', the mixing of the milled grist and the brewing liquor at the correct temperature and in the correct proportions to obtain the mash. After a period, with or without temperature changes, during which the necessary biochemical changes occur, the liquid 'sweet wort', which contains the extract, is separated from the residual solids, the 'spent grains' or 'draff'. Some extract remains in the draff, and as much of this as possible is recovered by 'sparging', washing the grains with hot brewing liquor.

Lautering

The operation of lauter tuns is often fully automated. The flow of wort and its cumulative volume, sparge liquor temperature, volume and flow, wort turbidity (haziness), the pressure differences across the grain bed (between the top of the vessel contents and the under-deck space), and between the under-deck space and the end of the collection pipe or the central collection vessel, and the positions of the knifing machinery (rotating speed, or stationary, height, direction of movement, knife setting, setting of grain discharge plough) are all determined and the measurements are fed to a computer. Increases in the cross-bed pressure instigate deeper cutting into the grain bed while increases in wort turbidity initiate reductions in the depth of cut. Because lautering is often the slowest process stage in wort production, there is a constant pressure to reduce wort separation times. The number of cycles that could be achieved with a lauter tun in 24h has risen from six to ten, with 12 being routinely achieved in some instances and even 15 cycles/24h being claimed. However, for these high rates of use to be achieved low bed loadings must be used.

Boiling

In traditional brewing, as practised in homes or small inns, hot water was placed in a wooden tub or tun, and the grist (malt that had been ground between millstones) was mixed and mashed in by stirring or rowing with a rake, paddle or 'oar'. No reliable means of measuring temperatures was available. In one method, which gave rise to the 'classical' British infusion system, the water temperature was guessed to be suitable by feel or by how clearly the brewer's face was reflected in the water. After a period a basket was pushed into the mash and wort that seeped into it was ladled into a receiver, in readiness for boiling with hops or other flavouring herbs. When wort recovery became difficult more hot water was mixed into the mash (re-mashing) and another, weaker wort was recovered. This was repeated until the worts were too weak to be worth collecting. In later times wort was collected from mashes using primitive mash tuns, in which the wort drained from the mash through a perforated 'strainer' in the base of the tun.

Adjuncts

Adjuncts are materials, other than malt, that are sources of extract. They are used because they yield less expensive extract than malt and/or they impart desirable characteristics to the product. For example, they may dilute the levels of soluble nitrogen and polyphenolic tannins in the wort, allowing the use of high-nitrogen (protein rich) malts and the production of beer less prone to form haze. Some adjuncts enhance head formation and retention. The higher proportion of adjuncts used in a mash the more difficult it is to achieve good extract recoveries and also wort viscosity is often increased, run-off is slowed and fermentability is reduced. The addition of soluble sugars or syrups to the wort effectively increases the capacity of the brewhouse and provides a simple method for generating high-gravity worts and adjusting wort fermentability. Solid, 'mash tun' adjuncts may be added to the grist and the starch they contain will be hydrolysed by enzymes from the malt or from other sources. Other soluble preparations, sugars and syrups, otherwise 'copper' or 'kettle' adjuncts, are dissolved in the wort during the hop-boil. In addition to these a brewer may add other sugars to the beer as 'primings', and caramels or other materials may be added to adjust beer colour.

Questions on Text

What is the function of malting?

Words to Watch

starch [stɑ:tʃ] n. 淀粉 vt. 给……上浆
corn grit 玉米粉
perlite ['pə:laɪt] n. 珍珠岩
sulfite ['sʌlfaɪt] n. 亚硫酸盐
corn [kɔ:n] n. （小麦等的）谷物；谷粒
formation [fɔ:'meɪʃən] n. 组成，形成；组成物，形成物
enzyme ['ɛnzaɪm] n. 酶
sugar ['ʃʊgə] n. 食糖；一匙糖；一块方糖；（植物、水果等所含的）糖
dosage ['dəʊsɪdʒ] n. （通常指药的）剂量
aeration [eɪə'reɪʃ(ə)n] n. 通风，充气
trub [t'rʌb] n. 凝固物，冷却残渣
artificially ['ɑ:ti'fiʃəli] adv. 人工地；不自然地
dry [draɪ] v. 把……弄干
mash [mæʃ] n. 糖化醪
whirlpool ['wɜ:lpu:l] n. 回旋槽；旋涡
amino acid n. 氨基酸
aldehyde ['ældɪhaɪd] n. 醛
ketone ['ki:təʊn] n. 酮类
organic acid 有机酸
by-product ['baɪˌprɒdʌkt] n. 副产品；副作用

Phrases and Patterns

contribute to 贡献

take up 开始从事；占用

be excreted from 从……排出

Supplementary reading

Reading Material 1

Beer is one of the most popular alcoholic beverages, being consumed in large amounts worldwide, which is a complex mixture containing numerous flavour-active volatile organic metabolites (VOMs) belonging to a diversity of chemical families over a wide range of concentrations, polarities and volatilities. These VOMs reflect the brewing process and have a strong influence on the quality and character of the beer. Consequently, they are also important for beer characterization. Overall, beer flavour results from a complex combination of different aspects that give each brew distinctive profiles. Such interferences result mainly from the ingredients composition, the roasting malt and boiling wort conditions, the metabolites produced by yeast during fermentation as well as the ones produced by contaminant microorganisms and also the effects of oxygen and sunlight during product storage. In generic terms, the brewing process involves four main steps, wort preparation, fermentation, maturation, and filtration and/or stabilization.

Reading Material 2

Traditionally, wort was boiled in direct-fired kettles, often made of copper. Since the heat source is localized at the bottom of the kettle, these vessels are not efficient in transferring heat into the wort, can scorch the wort, and are restricted by the volume of wort that can be boiled at any one time.

Lesson 5　Fermentation and Maturation

Objectives

At the end of this lesson, students should be able to:
1. Know what the role of primary fermentation is.
2. Know what the rate-limiting step is in the maturation of beer.
3. Know how diacetyl is formed.

Text

In brewing, the primary fermentation is one of the longest stages as well as an important aromatic compound production step. Indeed, fermentation has the main impact on process productivity and product quality. During the last fifty years, brewing process productivity has improved through better control of the various unit operations. In lager beer production, the fermentation time is decreased by increasing the temperature (from 8°C to 13~14°C), but to control the aroma profile characteristic of lager beer, top pressure is also used. Whereas temperature has a general accelerating effect, the combination of temperature and top pressure mainly control aromatic compound production. It is widely recognized that the real parameter affecting the production of volatiles is dissolved carbon dioxide, whose concentration depends on the temperature and top pressure applied.

The removal of diacetyl is the rate-limiting step in the maturation of beer. Diacetyl is formed by a nonenzymatic oxidative decarboxylation of α-acetolactate. In the subsequent stage of beer maturation, diacetyl is transformed to acetoin by yeast reductase. Since beer maturation very often is the bottleneck in beer production and ties up capital in the stored beer, technologies based on physical, biological, and biotechnological principles have been developed to accelerate beer maturation. The addition of the enzyme α-acetolactate decarboxylase to pitched wort is a simple and effective technology to limit the amount of diacetyl formed during fermentation and to reduce maturation time to a minimum. α-acetolactate decarboxylase transforms α-acetolactate directly to acetoin without the formation of diacetyl as an intermediate. Recently, the Food and Drug Administration approved the use of α-acetolactate decarboxylase for brewing applications in the United States. The properties of Maturex and the practical experience gained from brewing trials and production brewing are reviewed.

Adjuncts can be added during fermentation. Adjuncts are nothing more than unmalted grains such as corn, rice, rye, oats, barley, and wheat. They are readily available, other definite advantages are also achieved. Adjunct use results in beers with enhanced physical stability, superior chill-proof qualities, and greater brilliancy. Adjuncts can be used to adjust fermentability of a wort. Many brewers add sugar and/or syrup directly to the kettle as an effective way of adjusting fermentability, rather than trying to alter mash rest times and temperatures.

Adjuncts are often used for their flavor contribution. For example, rice has a very neutral aroma and taste, while corn tends to impart a fuller flavor to beer. Wheat tends to impart a dryness to beer. Semi-refined sugars add flavor to ales that has been described as imparting a luscious character. Adjuncts will also alter the carbohydrate and nitrogen ratio of the wort, thereby affecting for formation of byproducts, such as esters and higher alcohols.

Adjuncts are used for color adjustment, as in the case with dark sugars. On the other hand, adjuncts such as rice and pure starches and sugars are used to dilute malt colors to produce lighter colored beers.

Some adjuncts are used for their chemical properties; e. g., raw barley and wheat, which contribute glyco-proteins to enhance foam stability. Other adjuncts, low in protein, are used to improve colloidal stability since they will dilute the amount of potential haze-forming proteins.

Questions on Text

What does yeast reductase convert diacetyl into in the subsequent stages of beer maturation?

Words to Watch

osmolarity [ɒzməˈlærɪtɪ] n. 渗量
osmotic [ɒzˈmɒtɪk] adj. 渗透性的
sorbitol [ˈsɔːbɪtəl] n. 山梨醇
acidification power 酸化力
approximately [əˈprɒksɪmɪtlɪ] adv. 近似地，大约
parameter [pəˈræmɪtə] n. 参数，参量，变量
biotechnological [ˌbaɪəʊteknəˈlɒdʒɪkl] adj. 生物技术的
aromatic [ærəʊˈmætɪk] adj. 芳香的，有香味的
acetoin [əˈsetɒɪn] n. 乙偶姻
reductase [rɪˈdʌkteɪs] n. 还原酶
accelerate [əkˈsɛləreɪt] v.（使）加快
acetolactate [eɪstɒˈlækteɪt] n. 乙酰乳酸
decarboxylase [diːkɑːˈbɒksəleɪs] n. 脱羧酶

Phrases and Patterns

It is widely recognized that 众所周知
addition of 添加

Supplementary reading

α-acetolactate decarboxylase isolated from *Enterobacter aerogenes* strain 1033 has been applied for maturation of beer. Addition of the enzyme to freshly fermented beer effected removal of α-acetolactate and α-aceto-α-hydroxybutyrate in 24 hours at 10°C to a level below the taste threshold of the corresponding volatile diketones, diacetyl and 2,3-pentanedione, without affecting other important properties of the beer. By comparison of organoleptic properties, the beer matured in the presence of α-acetolactate decarboxylase was judged to be of an equally satisfactory quality when compared with conventionally prepared beer. The evidence suggest that apart from removal of diacetyl, 2,3-pentanedione and the precursors of these compounds from fermented beer no other important events related to flavour development occur during maturation of the type of beer studied. Consequently, it is concluded that application of enzymes may allow flavour maturation of beer to be carried out in less than 24 hours.

Lesson 6 Beer Flavor and Stability

Objectives

At the end of this lesson, students should be able to:
1. Know how to evaluate the flavor stability of beer.
2. Know how to accelerate the natural aging of beer.
3. Know what the main methods of forced aging are.

Text

Assessment of flavor stability of lager beer is typically done by gas chromatography (GC) and sensory analysis. When assessing sensory stability by GC, different methods are used to determine volatile organic compounds (VOC), each with their own advantages and disadvantages. The ideal method would indiscriminately extract all key aroma compounds without modifying any of the VOCs.

In 2009 Saison et al confined the list of indicators by re-determining the flavor thresholds of several compounds present in lager beer. They claimed that methional, 3-methylbutanal, 2-furfuryl ethyl ether, β-damascenone, and acetaldehyde, and to a lesser degree (E, E)-2, 4-decadienal, phenylacetaldehyde, 2-methylpropanal, diacetyl, and 5-hydroxymethylfurfural, were the key contributors to the aged flavor of beer. The Strecker aldehydes methional and phenylacetaldehyde are generally accepted as indicators for thermal stress, especially during wort boiling. The same is true for 2-furfural, which is thought to derive from xylose species. Nonetheless, this pathway was not confirmed for packaged beer but 3-deoxypentosone was found to be a direct precursor. In general, Maillard products tend to give caramel or cooked notes but, due to their high flavor thresholds, they seem to have little impact on total flavor. 2-furfuryl ethyl ether is particularly well known to increase in all beers during aging and has been suggested as a potent analytical storage indicator.

Maltodextrin can also affect the flavor of beer. It is the most complex fraction of the products of starch conversion. It is tasteless, gummy, and hard to dissolve. It is often said to add body (palate fullness) to beer, increase wort viscosity, and add smoothness to the palate of low-malt beers. However, it is easy to increase the dextrin content of grain beers by changing the mash schedule or using dextrin malt. Maltodextrin is of interest mainly as a supplement to extract brews.

Forced-aging (formerly "beer punishment") is a pervasive but discriminative way to accelerate the processes that occur during the natural aging of beer and thus predict flavor stability. Since brewers are not able to wait several weeks or months until the first perceivable changes occur, they depend on sensory and analytical tools for the rapid estimation of flavor stability in beer. Therefore, different forcing regimes are used involving changes in parameters such as temperature, time, mechanical action (e. g., shaking), impact of light, oxygen content (Table 1-3), and occasionally even alterations in pH value.

Table 1-3 Oxygen concentration after bottling-chemical characteristics beer samples

Avg. (STD)	Samples with higher initial oxygen content / (μg/L)		Samples with lower initial oxygen content/ (μg/L)	
TPO after bottling	1265 (185)		523 (382)	
HSO after bottling	1207 (180)		472 (380)	
DO after bottling	58 (44)		51 (38)	
Chemical characteristics	pH:	4.44	Eorg. (°P)	12.43
	Ale. (v/v%)	5.11	RDF (%)	63.71
	Ale. (w/w%)	4.00	ADF (%)	77.00
	Er (°P)	4.70	Cal. (kJ/100 mL)	188.19
	Ea (°P)	2.86		

Legend: TPO (Total Package Oxygen), HSO (Headspace Oxygen), DO (Dissolved Oxygen), pH (acidity), Ale. (Alcoholic volume or weight in %), Er (Real extract), Ea (Apparent extract), Eorg. (Original extract), RDF (Real degree of fermentation), ADF (Apparent degree of fermentation), Cal (Caloric content)

Questions on Text

What are the key factors that contribute to the aged flavor of beer?

Words to Watch

    gas chromatography 气相色谱分析
    volatile organic compound 挥发性有机物
    assessment [əˈsesmənt] n. 评估，评定
    volatile [ˈvɒlətaɪl] adj. 不稳定的；挥发性的
    organic [ɔːˈɡænɪk] adj. 有机的；不使用化肥的；有机物的；器官的
    compound [ˈkɒmpaʊnd] n. 化合物；复合物；混合物
    threshold [ˈθreʃəʊld] n. 阈值，界，临界点
    precursor [priˈkɜːsə(r)] n. 先驱；先锋；前身
    value [ˈvæljuː] n. （数学中的）值，（商品的）价值
    xylose [ˈzaɪləʊs] n. 木糖
    methional [ˈmeθənl] n. 甲硫代丙醛
    maltodextrin [ˌmæltəʊˈdekstrən] n. 麦芽糖糊精
    friability [fraɪəˈbilitɪ] n. 易碎性

Phrases and Patterns

    flavor stability 风味稳定性
    sensory analysis 感官分析
    be suggested as 建议为
    diastatic power 糖化力
    antiradical power 抗自由基能力
    reducing power 还原能力
    lipoxygenase activity 脂氧合酶活性
    nonenal potential 壬烯醛潜力

Unit 2　Wine

Lesson 1　The Variety of Grapes

Objectives

At the end of this lesson, students should be able to:
1. Know the differences between the red wine grape variety and white wine grape variety.
2. Know the main wine grapes in European wine region.

Text

Chardonnay

Chardonnay tends to produce fruit with high acid and high pH in cooler climates. In hot climates one can have problems with low acid.

Chardonnay produces medium bodied wine and is probably the only white variety with sufficient body to routinely accept wood aging. Some winemakers tend to go overboard with the wood content. Good examples can accept flavours derived from the primary fermentation lees and it can stand higher titratable acidities and alcohols than other white varieties.

Riesling

Riesling is described by English wine writer Jancis Robinson as of 'unbeatable quality; indisputably aristocratic. Ludicrously unfashionable'. This vine is very hardy and is grown in very cold climates because of its ability to survive winter temperatures of −20°C without damage to its wood. It is a moderately vigorous variety with a late budburst and ripening, has good disease resistance and produces reasonable crop yields with 10 tonnes/hectare easily achievable.

Cabernet Sauvignon

This noble red grape is one of the varieties responsible for the top Bordeaux clarets. It is only the newer wine regions of the world that make this wine as a pure varietal. This trend now seems to be loosing favour as more winemakers blend this variety with Merlot and more recently, Cabernet Franc.

The major criticism levelled at straight Cabernet Sauvignon wines is that they often have a "hole" in the middle of the palate (a reduction in sensation in the middle of the flavour profile). This lacking of richness and roundness can be rectified by the careful addition of Merlot. Merlot generally produces higher levels of sugar and this can be very useful in poorer years. Cabernet Franc adds a degree of elegance to the blend.

Merlot

This variety performs better on heavier soils than Cabernet Sauvignon and will ripen when Cabernet Sauvignon does not.

Viticulturally, poor set can be a problem and overcropping can be a problem in some years. Merlot has good disease resistance although botrytis can be found in cases of poor management.

Merlot produces full bodied red wines with cherry/plum characters but doesn't have quite the fruit intensity, tannins or colour of Cabernet Sauvignon.

Merlot wine is best blended with Cabernet Sauvignon at quite high levels. Cabernet Sauvignon and Merlot blends are generally better balanced than straight varietals made from either variety.

Pinot Noir

This grape is responsible for the great reds of Burgundy and is one of the main grapes used in making champagne. It has not however adapted as well to other regions of the world as has Cabernet Sauvignon.

Oenologically, this is a very difficult grape to come to terms with and very few great wines have been produced outside of France. This may in part be due to winemakers having not learnt to handle the variety fully. It takes considerably more manipulation from the winemakers to produce a great Pinot Noir than say a great Cabernet Sauvignon.

Sauvignon Blanc

Typically Sauvignon Blanc wines are recognized as not being long living. Initially, they start off tasting strongly of gooseberries. After about two years, this changes to what has been described as tinned peas and this remains typical for the next five or six years after which time, the wine starts to fade away and lose character. While these wines may taste fairly acidic when young, this sensation diminishes with age. Many winemakers blend up to 20% Semillon with Sauvignon Blanc. Semillon has a similar taste profile to Sauvignon Blanc however and can be used to fill out the middle palate and improve the wine. In Bordeaux, Semillon and Sauvignon Blanc are blended together to produce the traditional Sauternes and dry wines labelled Graves.

European wine region:

1. Bordeaux

Red grape variety: Cabernet Sauvignon、Merlot、Cabernet Franc、Petit Verdot、Malbec、Carmenere

White grape variety: Sauvignon Blanc、Semillon、Muscadelle

2. Burgundy

Red grape variety: Pinot Noir

White grape variety: Chardonnay、Chablis、Macon、Cote d'Or、Aligote

3. Loire Valley

White grape variety: Chenin Blanc、Melon de Bourgogne、Gamay、Chinon、Bourgueil

4. Italy

Barbera、Brunello、Dolcetto、Nebbiolo、Sangiovese

5. Spain

Tempranillo、Mourvedre、Grenache

Questions on Text

Which is the most famous white grape?

Words to Watch

Bordeaux [bɔːˈdəʊ] n. 波尔多（法国西南部港市）；波尔多葡萄酒

Burgundy [ˈbɜːgəndi] n. 勃艮第；勃艮第葡萄酒

Chardonnay [ˈʃɑːdəneɪ] n. 霞多丽（白）

Riesling [ˈriːslɪŋ] n. 雷司令（白）

Cabernet Sauvignon 赤霞珠（红）

Cabernet Franc 品丽珠；卡本内弗朗（红）

Merlot [ˈmɜːlət] n. 美乐（红）

Pinot Noir 黑比诺（红）

Sauvignon Blanc 长相思（白）

viticultural [vɪtɪˈkʌltʃərəl] adj. 葡萄栽培的

Lesson 2 Viticulture

Objectives

At the end of this lesson, students should be able to:
1. Know the importance of adapting viticulture to climate change.
2. Know what the viticultural practices are.
3. Know what the influencing factors of terroir are.

Text

It is a common belief that wines from organic viticulture are lower quality wines with respect to conventional ones, both as concerns sensory characteristics and a supposed higher content in compounds harmful for human health (e.g. ochratoxin A or biogenic amines). Nevertheless, these opinions remain mainly based on hypothetical considerations, because of the lack of scientific data reporting results on the analytical and sensory characterization of organic wines, as well as on the comparison between organic and conventional products.

Climate change is one of the most studied topics in the last decades due to its socio-economic, environmental and biological implications. It is expected that climate change will have a significant impact on various economic sectors but an especially large one on agriculture because crop yields are heavily influenced by weather conditions during their life cycle. In this context, perennial crop production, such as viticultural production, is considered highly vulnerable to climate change due to the fact that grapevine is extremely sensitive to climate and climate variability. In addition, viticultural production does not move in tandem with climate change due to many socio-economic factors such as long reestablishment periods, proximity to wineries, availability of labour and accessible markets, among others. Therefore, the importance of adapting viticulture to climate change lies on three fundamental aspects: (i) grapevines are planted for several decades, and new plantations may take 15~30 years to give full returns; thus, the selected cultivar should be adapted to a changing climate; (ii) the regulations on production techniques and varieties evolve slowly; and (iii) the qualitative characteristics of its final product is not only the result of terroir, which expresses the relationship among vines, ecopedology, the cultivation process and climatic variables, but also an expression of cultural and socio-economic parameters.

"Terroir" is a French term of wine industry. It embraces several meanings including the tradition, climate, soil, humidity, sunshine and all the elements like these making a wine-making region unique to the other ones all over the world. At its core is the assumption that the land from which the grapes are grown imparts a unique quality that is specific to that region.

Wine contains many metabolites originating from the grapes and from alcoholic and malolactic (ML) fermentations. In general, a wines metabolic profile could be affected by the "terroir", which accounts for factors of climate, land and cultural practices. The viticultural practices, such as soil management, canopy, irrigation together with winemaking technologies play the most important roles in making the overall, individual fingerprint of wine. Metabolomics in foods is very useful to account for the metabolic changes that occur during fermentation and/or production processes, and to evaluate the quality of food and beverages such as oil and wine.

Questions on Text

What are the susceptible influences during the viticulture production process?

Words to Watch

viticulture [ˈvɪtɪkʌltʃə] n. 葡萄栽培
conventional [kənˈvɛnʃənl] adj. 依照惯例的；传统的；常规的
harmful [ˈhɑːmfʊl] adj.（尤指对健康或环境）有害的，导致损害的
vulnerable [ˈvʌlnərəbl] adj. 脆弱的，易受伤害的
climate [ˈklaɪmət] n. 气候；风气
variability [ˌveərɪəˈbɪləti] n. 可变性；易变性；反复不定
metabolic [ˌmetəˈbɒlɪk] adj. 新陈代谢的
fingerprint [ˈfɪŋɡəprɪnt] n. 指纹；指印
evaluate [ɪˈvæljʊeɪt] vt. 估计；评价；评估

Phrases and Patterns

It is a common belief that 人们普遍认为
be affected by 受……影响

Supplementary reading

Climate change is expected to impact considerably on viticultural zoning. In this work, impacts of future climate on Argentinean winegrowing regions are assessed using climatic projections from the IPSL-CM5A-MR model for the near (2015-2039) and the far (2075-2099) future under two emission scenarios (RCP4.5, RCP8.5). Four bioclimatic indices were assessed for exploring possible geographical shifts and suitability changes of Argentinean winegrowing regions. These geographical and suitability changes, assessed in terms of changes in vineyards location, varieties selection, and quality and quantity of grapevines, were considered as challenges and/or opportunities depending on the winegrowing region, the temporal horizon and the emission scenario. Results show a significant southwestward and higher altitude displacement of winegrowing regions, mainly for 2075-2099 under RCP8.5 scenario. Accordingly, the Argentinean viticulture may face both opportunities and/or challenges due to projected warmer climate conditions. Winegrowing regions of cold climates could be favoured, while winegrowing regions of warm climates could be disadvantaged, mainly for 2075-2099 under RCP8.5. Therefore, warmer climate conditions might be beneficial for maintaining the current grapevine varieties with their current grape quality or for cultivating new grapevine varieties in the new projected winegrowing areas; or they might be harmful for maintaining the current grapevine varieties with their yields and quality over most of current grapevine growing areas, but especially in the projected warmest ones. Summarizing the results allow for understanding how temperature and precipitation changes could affect the future geographical distribution of Argentinean winegrowing areas and their suitability characteristics, useful information for planning adaptation in the coming decades.

Adventitious roots

Roots that develop in areas of the grapevine where there was previously no root system, such as the roots that develop from the nodes of a newly planted cutting. While grapevines have adventitious roots, they do not have adventitious buds and requiring pre-existing buds for future growth.

Anthocyanins

Polyphenols located in the skin of grapes that includes the colour pigments that gives both grapes and wine their colour.

Balance pruning

A method of pruning based on the amount of growth that the vine experienced the previous growing season. This is often determined by weighing the one-year-old that is pruned during the winter dormancy period and using a formula to determine how many buds should be left for the next season's crop.

Cane pruning

Pruning method where the one or two canes of 1 year old wood is left on the vine after winter pruning with between 8 to 15 buds.

Irrigation

The supplementation of water in the vineyard either by drip-systems, overhead sprinklers or canals. While commonly used in New World wine regions, the practice was, until recently, banned in most wine-regions in the European Union.

Noble rot

Another name for the *Botrytis cinerea* mould that can pierce grape skins causing dehydration. The resulting grapes produce a highly prized sweet wine, generally dessert wine.

Vine density

The number of vines per a define area of land (acres, hectare, etc.). This can be influenced by many factors including appellation law, the availability of water and soil fertility and the need for mechanization in the vineyard. In many wine regions vine density will vary from 3000 to 10000 vines per hectare.

Yield

In any farming capacity, the quantity of quality fruit that a parcel of land render after a harvest. In terms of wine making it is the quantity of grapes that a vineyard can produce per hectare (2.47 acres) of land to produce the level of quality desired.

Lesson 3 Red Wine Vinification

Objectives

At the end of this lesson, students should be able to:
1. Describe appropriate fermentation methods for the common red wine.
2. Discuss the advantages and disadvantages of conventional red fermentations.

Text

The main feature of red wine fermentation is impregnation fermentation. That is, in the fermentation process of red wine, alcohol fermentation and the impregnation of solid matter coexist. The former converts sugar into alcohol, and the latter dissolves phenolic substances such as tannins and pigments in the solid matter in the wine.

①Stalk removal and crushing: grapes—vibration screening table to remove impurities and small green grains—movable lifting frame—stalk removal and crushing machine to remove fruit stems and crush them—the juice trough and pulp pump collect and transport the crushed pulp to the fermentation tank.

②Canning: Immediately when the grapes are crushed and de-stemmed and then pumped into the fermentation tank, and SO_2 is added while the can is filled. After filling, a backfill is performed to mix the SO_2 and the fermentation substrate evenly. The amount of addition depends on the hygienic status of the grapes, generally $50 \sim 80 mg/L$. Certain potential spoilage organisms that are sensitive to sulfur dioxide, can grow if no sulfur dioxide is added at this stage. Pectinase can decompose on grape skin and promote the impregnation process of pigment, aroma and tannin. Although SO_2 has less effect on pectinase, it should be avoided at the same time. Add $20 \sim 40 mg/L$.

③Add yeast: put dry yeast in the ratio of 1: (10~20) into warm water at $36 \sim 38℃$ for 15~20 minutes, or activate in 2%~4% sugar water for 30~90 minutes to make yeast emulsion, then added to the mixture for fermentation. After the yeast is added, a beating cycle is performed to mix the yeast and the fermented mash evenly.

④Fermentation process: monitor the fermentation temperature, control the fermentation temperature at $25 \sim 30℃$, measure the specific gravity every 4~6h, and record it with the temperature in the original wine fermentation record table. $28 \sim 30℃$ is good for brewing wines with high tannin content and long aging time, while $25 \sim 27℃$ is suitable for brewing fresh wines with strong fruit flavor and relatively low tannin content. "Cap", the temperature of the fermentation substrate rises. If the quality of the raw materials is not good, to achieve a certain degree of alcohol, a certain amount of sugar needs to be added after fermentation enters vigorousness. Perform backfilling and spraying. The number of backfilling depends on many factors, such as the type of wine, raw material quality and dipping time, etc. Generally, the backfilling is performed 1~2 times a day, about 1/3 each time. This process generally lasts about 1 week.

⑤Separation and squeezing of skin residue: When the specific gravity of wine is reduced to 1000 and below (or the sugar content is determined to be less than 2g/L), the separation of skin residue starts. After separation, in order to ensure the progress of alcohol fermentation, the temperature of the free-flowing wine should be controlled at 18~20℃, full tank.

⑥ Malic acid-lactic acid fermentation: Malic acid-lactic acid fermentation is a necessary process to improve the quality of red wine. Only after the malolactic fermentation is completed and proper SO_2 treatment is carried out, the red wine has biological stability. And the wine becomes softer and more rounded. This fermentation process must be guaranteed to be full and sealed. After the end, add SO_2 to 50mg/L (If red grapes are processed with detectable spoilage, sulfur dioxide should be added. The greater the degree of rot, the higher the level of sulfur dioxide, however it should not exceed 60~70 mg/L, otherwise the malolactic fermentation may be inhibited.).

Pectin-splitting enzymes are generally not added. Some winemakers do not add sulfur dioxide to clean fruit when processing Cabernet Sauvignon, Merlot, Cabernet Franc and similar varieties. With Pinot Noir, there is no general consensus of opinion whether sulfur dioxide should be added and if so how much.

Questions on Text

What are appropriate fermentation methods for the common red wine?

Words to Watch

phenolic [fɪˈnɒlɪk] adj. 酚类的
vibration screening 振动筛选
pigment [ˈpɪgmənt] n. 色素；颜料
pectinase [pekˈtiːnəz] n. 果胶酶
yeast [jiːst] n. 酵母
softer [ˈsɒftə] adj. 温和的，柔软的
stalk removal 除茎
crush [krʌʃ] v. 破碎，压碎
de-stemmed [diːstɛmd] vt. 除梗

Supplementary reading

With most varieties of red grapes, the juice is colourless and the colour is associated with the skins. The process of red winemaking is principally involved in using the alcohol produced in the fermentation to leach out the colour from the skins. Particular care must be taken to maximise colour extraction from varieties such as Pinot Noir which tend to produce low levels of tannins and anthocyanins (compounds responsible for red colour in red wines). Much of the colour extracted during the fermentation process is unstable and drops out at the end of fermentation or in the first few months following fermentation. Winemakers use techniques to try to minimise this colour loss.

Colour extraction and fermentation techniques can be broadly grouped into conventional fermentations in open or closed tanks, carbonic maceration and heat extraction. Conventional fermentation in open or closed tanks would account for the vast bulk of red wines produced (Figure 1-1).

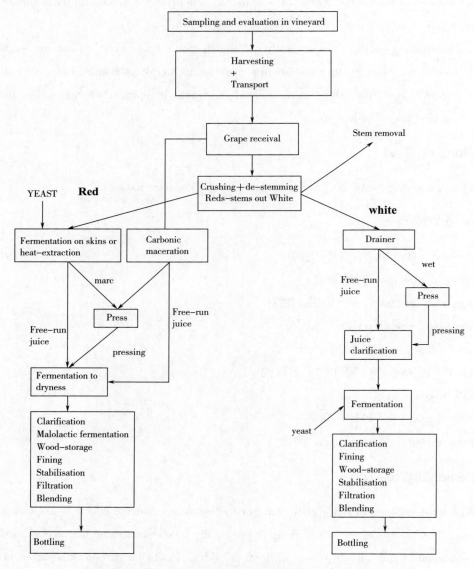

Figure 1-1 Flow diagram for red and white table wine production

In the French literature, the terms cuvaison and maceration are used when discussing red wine fermentations. Cuvaison relates to the fermentation process. Maceration as related to red wine production means the mixing of skins and juice and/or wine to extract colour. Up to this stage, the term maceration has been used to describe any process leading to the production of damaged grapes and grape solids.

Related concepts

Aerobe

An organism requiring molecular oxygen (O_2) for growth.

Aerobic respiration

Respiration in which the final electron acceptor in the electron transport chain is molecular oxygen (O_2).

Anabolism

All synthesis reactions in a living organism; the building of complex organic compounds from simpler ones.

Anaerobic respiration

Respiration in which the final electron acceptor in the electron transport chain is an inorganic molecule other than molecular oxygen (O_2), e.g., a nitrate ion or CO_2.

Ascospore

A sexual fungal spore produced in an ascus, formed by the ascomycetes.

Budding

A sexual reproduction beginning as a protuberance from the parent/mother cell which becomes a daughter cell.

Carbohydrate

An organic compound composed of carbon, hydrogen, and oxygen; carbohydrates include starches, sugar and cellulose.

Chemically defined medium

A culture medium in which the exact chemical composition is known.

Citric acid cycle

A metabolic pathway that concerts two-carbon compounds to CO_2, transferring electrons to NAD^+ and other carriers; also called Krebs cycle or tricarboxylic acid (TCA) cycle.

Lesson 4 White Wine Vinification

Objectives

At the end of this lesson, students should be able to:
1. Describe appropriate fermentation methods for the common white wine.
2. Discuss the advantages and disadvantages of conventional white fermentations.

Text

White wine is an alcoholic beverage obtained by alcohol fermentation of white grape juice. During the fermentation process, there is no impregnation phenomenon of grape juice on the solid part of grapes. The quality of dry white wine is mainly determined by the first-class aroma of grape varieties, the second-class aroma derived from alcohol fermentation and the content of phenolic substances. Therefore, under certain conditions of grape varieties, the extraction speed and quality of grape juice, factors affecting the formation of second-class aroma, and the oxidation of grape juice and wine have become important factors affecting the quality of dry white wine (Figure 1-2).

①Stalk removal and crushing: grapes—vibration screening table to remove impurities and small green grains—movable lifting frame—stalk (or stem removal only) removal and crushing machine to remove fruit stems and crush them—the juice trough and pulp pump collect and transport the crushed pulp to fermentor.

②Pressing and extracting juice: the airbag and the tank wall only squeeze the material during pressing, and the friction effect is very small, and it is not easy to squeeze out the peel, fruit stems and components of the juice in the province, so the content of solid matter and other undesirable components in the juice less.

③Low temperature clarification and separation of clear juice: After the juice enters the heat preservation tank, add $60 \sim 120 mg/L$ of SO_2 and circulate evenly. In order to speed up the clarification and impregnation effect, clarification pectinase and bentonite and other lower leg materials can be added for the next gum treatment. Keeping the temperature at $0 \sim 5℃$, $24 \sim 48h$ low temperature can also impregnate more elegant aroma, and can control the amount of tannin leaching.

④Alcohol fermentation: The separated sake is quickly raised to $18 \sim 20℃$, and the yeast for liquor is added to start fermentation. Before the fermentation is started, the juice is taken to test various physical and chemical indicators. Sensory and physical and chemical analysis are carried out at any time during the fermentation process. Issues to be noted: 1. The canned tank should be full. 2. Temperature: 18~20℃, fill out the fermentation record form.

⑤Clarification and separation: clarify after fermentation. The separated sake is led to a storage tank, and SO_2 is added to 60 mg/L, which is sealed and stored. Pressed wine can be processed separately or mixed with sake. If the acidity is too high, consider malolactic fermentation. Add SO_2 after the fermentation to seal the storage. Hand Harvested White Grapes that are to be Crushed Grapes coming into the winery in bins are generally tipped into the destemmer-crusher with a bin tipper.

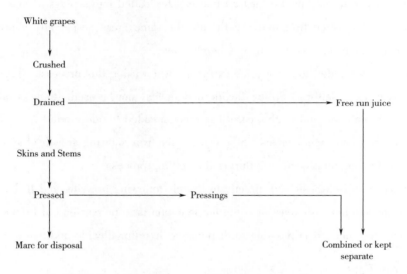

Figure 1-2　Flow Chart for White Grape Handling in the Winery

If the grapes come into the winery in picking boxes, the boxes are tipped directly into the destemmer-crusher. As the throughput in this instance is generally fairly slow, smaller destemmer-crushers with a capacity of four tonnes per hour can be used. The destemmer-crusher can often be mounted directly above a drainer (if used), or directly over a press. This does away with the need for an expensive must pump and also is much more gentle on the grapes.

Some wineries mount a sorting belt prior to the crusher and unload the grapes onto this for final sorting prior to destemming and crushing. This is useful when processing grapes with mixed maturity or with a high level of rot. Sorting grapes on a sorting belt is preferable to asking pickers to leave diseased fruit on the vines. However, unripe grapes can be left and harvested later if there are sufficient to warrant this.

The same comments apply to the use of sulfur dioxide and enzyme additions as for mechanically harvested fruit.

Hand Harvest White Fruit for Whole Bunch Processing

In this case, the fruit is loaded directly into the press. This technique is used where it is desired to maximise the quantity of top quality free run juice which is low in phenolics and colour. This technique is used for processing juices for bottle fermented sparkling wines and for some top quality Chardonnay.

Some wineries also use this technique for machine harvested fruit. The perceived advantage in this instance is the reduction of the mechanical damage incurred in harvesting and handling.

A good quality batch press needs to be used for this technique. The juice extraction is slow. The quantity of whole bunches of fruit that can be loaded into a press is considerably less than the quantity of crushed fruit that can fit into the same press, especially if juice is being drained off at the same time as the press is being filled.

Enzymes are not added to assist juice extraction when using this process. Enzymes may be added to assist in juice settling. Sulfur dioxide may or may not be used. Juice is normally cold settled before fermentation and in this case it is recommended to add around 25mg/L sulfur dioxide for clean fruit and higher levels (50+ mg/L) for fruit showing growth of spoilage organisms to prevent fermentation occurring during the settling process.

If the juice is to be transferred directly to barrels following pressing and if this is followed by immediate inoculation with yeast or if the fermentation is to be carried out by the yeast naturally present in the juice, it is possible to dispense with sulfur dioxide addition at this point.

Questions on Text

What are appropriate fermentation methods for the common white wine?

Words to Watch

airbag ['eəbæg] n. 气囊
dry white wine 干白葡萄酒
clarification [ˌklærɪfɪ'keɪʃən] n. 澄清
separation [ˌsɛpə'reɪʃən] n. 分离
bentonite ['bentəˌnaɪt] n. 膨润土
leach [li:tʃ] v. 过滤

Supplementary reading

Many different methods are used for processing white grapes into wine. These methods vary from winemaker to winemaker, variety to variety and country to country.

The aim in processing these varieties is to retain the distinct floral or spicy aromas found in the fruit. These characters can be quite delicate and the processing should aim to maximise the retention of these aromas and avoid extracting other flavours or aromas that could mask them. Care should be taken to avoid oxidation.

Following harvest, the fruit is crushed, drained and pressed. While most winemakers feel that skin contact before draining is desirable, care has to be taken to avoid extracting harsh tannins at the same time. This is especially true with Riesling and Gewurztraminer.

The juice is cold settled, racked and inoculated with either a neutral or aromatic yeast. If it is intended to stop ferment the juice a cold sensitive yeast should be used.

If the acid level is higher than considered desirable, the juice may be chemically deacidified. Malolactic fermentation is not normally employed.

The juice is fermented cold at around 1 Brix per day to conserve the fruit aromas. The fermentation may be stopped or allowed to proceed to dryness.

Neutral or aromatic yeast are used.

The young wine is then rapidly racked off gross lees to prevent the pick up of yeast autolysis characters.

These wines are then normally fined, cold stabilised, (wines contain high levels of potassium bitartrate and if the wine is not cooled to just above its freezing point and racked or filtered off the crystals formed, the wine is likely to produce more crystals in the bottle) filtered and bottled within 6~9 months of harvest.

Techniques such as whole bunch pressing, fermentation or ageing in oak and extended lees contact, are not normally practised.

Related concepts

Cytoplasm

In a prokaryotic cell, everything inside the plasma membrane. In a eukaryotic cell, everything inside the plasma membrane and external to the nucleus.

Deoxyribonucleic acid (DNA)

The nucleic acid of genetic material in all cells and some viruses.

Ecology

The study of the interrelationships between organisms and their environment.

Electron transport chain/system

A series of compounds that transfer electrons from one compound to another, generating ATP by oxidative phosphorylation.

Eukaryotic species

A group of closely-related organisms that can interbreed.

Glycolysis

The main pathway for the oxidation of glucose to pyruvate.

Pentose-phosphate pathway

A metabolic pathway that can occur simultaneously with glycolysis to produce pentoses and NADH without ATP production.

Sanitization

The removal of microbes from food and beverage preparation areas.

Lesson 5   Wine Chemistry

Objectives

At the end of this lesson, students should be able to:

1. Know what the main role of malic acid is.

2. Know what the purpose of malolactic fermentation (MLF) through the decarboxylation of L-malic acid to L-lactic acid is.

3. Know the significance of malolactic fermentation (MLF).

Text

During winemaking, the initial conversion of grape must to wine is an alcoholic fermentation (AF) carried out by one or more strains of yeast, typically *Saccharomyces cerevisiae*. After the alcoholic or primary fermentation, a secondary fermentation known as malolactic fermentation (MLF) is often undertaken, depending on the style of wine that the winemaker seeks to achieve. Malic acid is one of the predominant organic acid in grapes, occurring in amounts of the order of 3g/L. This compound can therefore contribute to the acidity, pH and mouthfeel of wine and can be a nutrient for several spoilage organisms. Accordingly, a malolactic fermentation (MLF) by which L-malic acid is decarboxylated to L-lactic acid serves several purposes: to reduce the harsh acidity of malic acid, give a concomitant modest increase in pH as well as increase wine microbial stability. The former of these outcomes are of greatest importance in sparkling and some white wines, whereas most red wines undergo MLF primarily for stability. Subsequent effects due to MLF also include impacts on both aroma and visual profile.

Malolactic fermentation (MLF) consists of the decarboxylation of malic acid into lactic acid and CO_2, and it is a required step for red winemaking because it is a natural and biological deacidification of wine that increases its stability, allows wine ageing and increases its positive sensory properties and aroma complexity. This secondary fermentation is carried out by lactic acid bacteria (LAB), mainly *Oenococcus oeni* (Ribéreau-Gayon, Dubourdieu, Donèche & Lonvaud, 2006). MLF can occur spontaneously in wine by the indigenous LAB strains. An initial population of $106 \sim 107$ CFU/ml is considered appropriate for initiating malic acid degradation, and physical and chemical characteristics of the wine (temperature, pH, ethanol content, etc.) are also crucial for MLF to take place. Long-established winemaking protocols involve inoculation of selected LAB starter cultures to initiate and conduct the MLF of the inoculated wine. This inoculation of selected LAB starters provides an efficient control of wine MLF, nevertheless, both inoculated and spontaneous MLFs should be monitored to prevent LAB overgrowing and disturbing the harmonious equilibrium of wine components. (White wine should be clarified as quickly as possible after the cessation of fermentation except where it is intended to produce complex white wines that will undergo a malolactic fermentation or where extended lees contact character is desired).

Questions on Text

What is the main organic acid in wine?

Words to Watch

alcoholic [ˌælkəˈhɒlɪk] adj. 酒精的
primary fermentation 初次发酵，主发酵
acidity [əˈsɪdɪti] n. 酸味；酸性
spoilage [ˈspɔɪlɪdʒ] n. 有害，变质
organism [ˈɔːgənɪzm] n. 有机体；生物；有机体系
degradation [ˌdɛgrəˈdeɪʃən] n. 降解，退化
malolactic fermentation (MLF) 苹果酸-乳酸发酵
lactic acid bacteria (LAB) 乳酸菌
alcoholic fermentation (AF) 酒精发酵
decarboxylate [diːkɑːˈbɔksəleit] vt. 脱羧基
stability [stəˈbɪlɪti] n. 稳定性
deacidification [diːəsidifiˈkeiʃən] n. 脱酸（作用）
inoculation [ɪˌnɒkjʊˈleɪʃən] n. 接种

harmonious [hɑː'məʊnɪəs] adj. 和谐的

equilibrium [ˌiːkwɪ'lɪbrɪəm] n. 均衡

Phrases and Patterns

be known as 被称为

impact on 对……的影响

depend on 依靠

Supplementary reading

The perceived aroma of a wine is generally complex, and is the result of aroma molecules of different origins; aroma compounds found in the grape, aroma compounds produced during grape processing (chemical, thermal and enzymatic reactions in the must), during the alcoholic fermentation (fermentation aroma) and during the maturation of the wine (maturation aroma). Aromas are perceived by humans because they reach the olfactory epithelium and a prerequisite for this that the component is volatile. The volatility of aroma compounds depends on the compounds' chemical structure and interaction with the media from which they are liberated, as well as other factors such as temperature.

To determine the phenolic maturity degree objective methods exist, corresponding to chemical analysis, that relate to the measurement of indicators such as pH, acidity, soluble solids, among others. These methods are generally very accurate, but involve laboratory analysis.

There are also subjective methods, which are based on the experience of the oenologist, and phenolic maturity is estimated by inspecting the aroma, flavor and appearance of both fruit and wine. These types of analysis are called organoleptic analysis.

Traditionally, organoleptic analyses are performed on the fruit of the grape, or the wine. Unlike previous work, a very recent line of research explores the seed of the fruit to determine the Phenolic Maturity. Such indicators as color, shape and texture of the seeds were studied, finding a high correlation between appearance indicators and the taste of wines.

Lesson 6 Wine Sensory Properties

Objectives

At the end of this lesson, students should be able to:

1. Know what sensory properties include.
2. Know what the characteristics of Chardonnay wine are.

Text

All sensory properties including aroma, taste, flavor-by-mouth, mouth-feel and color play key roles in defining the character and sensory profile of a wine. All varietal wines should preferably possess a distinct set of aroma attributes which defines those wines. Chardonnay wines (Chardonnay tends to produce fruit with high acid and high pH in cooler climates. In hot climates one can have problems with low acid.) have been found to have tropical fruit, floral and oaky aromas (De La Presa-Owens and Noble 1997; Lee and Noble 2003), although Cabernet Sauvignon wines (This noble red grape is one of the varieties responsible for the top Bordeaux clarets. It is only the newer wine regions of the world that make this wine as a pure varietal.) tend to have dark berry, cooked fruit, bell pepper and wood characters. Sensory descriptive analysis (DA) has been employed to describe the attributes of wines based on their variety and appellation. Using statistical methods such as principal component analysis (PCA), canonical variate analysis (CVA) and analysis of variance (ANOVA), the DA data obtained can be more easily explained and interpreted.

When a wine reduction is made, the process of aroma liberation is sped up, and aroma is liberated into the air due to elevated temperature and the effect of water evaporation. Correspondingly, in chemical terms, one can imagine that the impact of the reduction process is a loss of volatile components, i.e. aroma compounds, accompanied by increased concentration of the non-volatiles, i.e. taste components. Therefore, by reducing wines, one can imagine the aroma profiles to become more similar, due to significant evaporation of volatile components. On the other hand, the profiles of non-volatile components may diverge as their concentrations are increased due to the evaporation of water, ethanol and other volatile compounds. Finally, the effect of chemical reactions taking place while heating, represents an unknown factor, influencing the flavor of the reduction in an unknown direction. The basic question is thus whether the reduction process enhances differences and variation, or diminishes the variation of the overall sensory characteristics. Limited research work has previously been performed in the area of wine reductions.

Wine is highly complex, and as such, compositional analyses cannot reliably predict wine sensory profiles. Sensory studies therefore continue to provide valuable information when determining the outcome of treatments on wine aroma, flavor, taste and mouthfeel properties. Amongst the various sensory methods that are available, descriptive analysis (DA) is frequently used to characterize and/or contrast wine sensory profiles. However, few studies to date have employed descriptive analysis to profile the sensory properties of reduced alcohol wines (RAW). A key aim of this study was therefore to identify changes in the chemical and sensory profiles of five Cabernet Sauvignon wines, following partial dealcoholization by ROEP (reverse osmosis-evaporative perstraction) treatment.

Questions on Text

What is used to characterize and compare the sensory properties of wine?

Words to Watch

attribute [ˈætrɪbjuːt] n. 属性，特质；[əˈtrɪbjuːt] v. 把……归因于
oaky aromas 橡木桶香气
sensory [ˈsɛnsəri] adj. 感觉的；感官的
tropical fruit 热带水果
berry [ˈbɛri] n. 浆果；莓
enhance [ɪnˈhɑːns] vt. 提高；增强；增进
descriptive analysis (DA) 描述性统计分析
principal component analysis (PCA) 主成分分析
canonical variate analysis (CVA) 典型变量分析
analysis of variance (ANOVA) 方差分析
evaporation [ɪˌvæpəˈreɪʃən] n. 蒸发

Phrases and Patterns

be employed to 被用于
continue to 继续
take place 发生

Supplementary reading

In red wine, specifically Cabernet Sauvignon, the compound usually responsible for the bell pepper aroma is 2-methoxy-3-isobutylpyrazine (MIBP). Gas chromatography-mass spectrometry (GC-MS) methods have been used to quantify the amounts of MIBP found in wine. Further studies have linked measured levels of MIBP in wines to the sensory perception of bell pepper aroma in the same wines. Results from such chemical and sensory analyses were found to be highly correlated.

Brettanomyces

A yeast that can grow in finished wine. Produces compounds that in small amounts can add complexity to wine but can be considered to make a wine faulty.

Cork taint

A wine spoiled with TCA (trichloroanisole) is considered to be faulty.

Hydrogen Sulphide

A chemical produced by yeast during fermentation mainly when they are under stress and lacking nutrients. Notable for the smell of rotten eggs.

Mercaptan

The odour of methyl and ethyl sulphides often described as burnt rubber, rotten cabbage or rotten onion.

Volatility

The perceptible presence of acetic acid and ethyl acetate in wine.

Wine fault

A perceptible character in a wine that reduces the organoleptic quality.

Lesson 7　Sparkling Wine

Objectives

At the end of this lesson, students should be able to:
1. Know the traditional method of sparkling wine production.
2. Know where the aroma compounds in sparkling wine come from.

Text

Currently, the wine whose production has increased the most is the sparkling wine (OIV. The International Organization of Vine and Wine, 2017). Thus, in the last 10 years, sparkling wine production has increased above 40% instead still wine a 7%. This is partly due to the change of the trends of consumption, from mainly festive consumption to more regular consumption (OIV. The International Organization of Vine and Wine, 2017).

The traditional method of sparkling wine production consists of two fermentations; the first one to produce the base wine from the must, the second one in the bottle, and it is followed by a period of aging on lees. For the second fermentation, a tirage liqueur which contains saccharose and yeast starter is added to each bottle, to produce the required amount of CO_2 which is between 5 and 6 bars. During the aging on lees period, the bottles remain in horizontal position for maximum wine—sediment contact. This period is ended by riddling process, which helps to collect the yeast sediments on the neck of the bottle, before their removal.

Aroma is one of the most important indicators of sparkling wine quality. The results showed that the responsibility for fruity/floral nuances in sparkling wine might reside in a few high-impact aromatic compounds, such as ethyl isobutyrate, isoamyl acetate, ethyl hexanoate, β-phenylethanol and diethyl succinate. In particular, aroma compounds can have different origins: from the grape (pre-fermentative aroma), from the yeast during the first or second fermentation (fermentative aroma); or from ageing during settling (post-fermentative aroma). It seems that tertiary and secondary aromas determine differences between wines. There is an important loss of esters during the second fermentation and posterior aging period, which is related to a decrease in the contribution of the odoriferous zones related to fruity aromas. This loss might be due to adsorption onto lees, but also due to the chemical hydrolysis due to their thermodynamical unstability. Furthermore, also the liqueur d'expedition (added to the wine after the disgorging) can provides sweetness and a unique flavour to the wine.

Questions on Text

What is one of the most important indicators of sparkling wine quality?

Words to Watch

sparkling wine   n. 起泡葡萄酒
must [mʌst] n. 待发酵葡萄汁
lees [li:z] n. 酒脚，酒泥，酒底沉淀物
bottle [ˈbɒtl] n. 瓶子
liqueur [lɪˈkjʊə] n. 利口酒，香甜酒
saccharose [ˈsækərəʊs] n. 蔗糖
yeast starter   n. 酵母启发剂
horizontal [ˌhɒrɪˈzɒntl] adj. 水平的；横向的
sweetness [ˈswi:tnəs] n. 甜；芬芳
removal [rɪˈmu:vəl] n. 去除；移动
sediment [ˈsɛdɪmənt] n. 沉淀物；沉积物

Phrases and Patterns

be added to 添加到

it seems that 看起来

Supplementary reading

The usage of alternative non-*Saccharomyces* yeasts might provide desirable characteristics to white and red sparkling wines. Second fermentation in the bottle was carried out by *Saccharomyces cerevisiae* as control and two non-*Saccharomyces* species: *Saccharomycodes ludwigii* and *Schizo saccharomyces pombe*. The second fermentations of white base wine made from Vitis Vinifera cv. Airén grapes and red base wine made from Vitis Vinifera cv. Tempranillo grapes, in the bottle were followed by aging on lees during 4 months at 12℃. Finally, physicochemical properties were analyzed and a sensory evaluation was held. Significant differences were detected among sparkling wines produced with the studied yeasts in acidity parameters and non-volatile compounds. The pyranoanthocyanin content and color intensity was higher with the use of *Schizo saccharomyces pombe* in red sparkling wines. The total amount of volatile compounds was similar among treatments, but in certain compounds, individual variations in concentration were seen. Total amount of biogenic amines decreased in all the samples after the treatment. Differences were also detected in sensory evaluation; the sparkling wines produced with *Saccharomyces cerevisiae* showed different aromatic profile in comparison to sparkling wines produced with *Schizo saccharomyces pombe*, considering the parameters of limpidity, aroma intensity, aroma quality, flowery, fruity, buttery and reduction aromas in white samples; color intensity, limpidity, aroma intensity, herbal, buttery, yeasty, and oxidation aromas in red samples. Usage of non-*Saccharomyces* yeasts for sparkling wine production with traditional method can be further studied to change specific characteristics of sparkling wines without decreasing their overall quality and obtain differentiation.

Unit 3　Rice Wine

Lesson 1　Sticky Rice

Objectives

At the end of this lesson, students should be able to:
1. Know what the Chinese rice wine is.
2. Know how Chinese rice wine is brewed.
3. Know what the characteristics of Chinese rice wine are.

Text

Chinese rice wine (CRW), a national unique and traditional alcoholic beverage, has more than 4000 years of history and is very popular in China, especially in Southern China. CRW has the characteristics of a yellow color, a sweet aroma and abundant nutrition and is called 'liquid cake'. CRW is typically fermented from sticky rice, with yeast and with wheat Qu, by spontaneous fermentation in local factories in the summer, while CRW is made throughout the year. As a crude enzyme preparation, wheat Qu plays an important role in the diastatic fermentation, which is similar to the koji of sake. At the same time, wheat Qu is a raw material in wine-making and brings to the wine a unique flavor. In general, flavor compounds are mainly produced, but are not limited to yeast fermentation during CRW brewing. More than 2×10^6 microbe colonies per gram of wheat Qu are typically inoculated into the CRW fermentation broth. These microorganisms take part in the fermentation in addition to the yeast, and together they synthesize compounds. In long-term production practice, it is recognized that in addition to the brewing raw materials and process, the wheat Qu made in the local factories has a great impact on the flavor of Chinese rice wine, with CRW produced in different areas having distinct flavors.

Chinese rice wine, with its unique flavor and high nutritional value, is one of the oldest drinks in the world, having been consumed by Chinese people for many centuries. As a traditional alcoholic beverage in China, it is mainly divided into the rice wines Zhepai, Haipai and

Minpai based on local characteristics. With the growth of younger consumers, Haipai rice wine has become more and more popular for its refreshing taste. Haipai rice wine, one of the most representative types of Chinese rice wine, is brewed with purebred *Saccharomyces cerevisiae* and new craftsmanship, which has not only shortened the fermentation period but also has greatly reduced the probability of spoilage. However, purebred fermentation results in a less mellow flavor and taste because the structure of the microbial community is relatively simple. With both new and traditional brewing technologies, there are issues of instability in the dominant strains and the brewing process, which have more or less restricted the development of Haipai rice wine in China. Thus, the study of Chinese rice wine fermented with mixed yeast starter is of primary importance to alter the insufficient aroma and lighter body that result from pure fermentation in this wine (Table 1-4).

Table 1-4   Analysis of chemistry in Chinese rice wine

parameter	Concentration						significance
	Control	1.5 Glu	2Glu	1.5 Alb	2 Alb	1.5Glu+1.5Alb	
Total sugar/ (g/L)	2.2±0.2ab	22±0.3ab	1.7±0.1 b	2.2±0.2ab	2.6±0.2a	2.0±0.1 ab	*
Titratable acidity/ (g/L)	6.5±0.6 dab	6.7±0.5ab	6.9±0.4a	6.3±0.2b	6.3±0.4b	6.6±0.5ab	*
pH	3.6±0.3	3.7±0.4	3.7±0.3	3.6±0.3	3.5±0.3	3.7±0.5	NS
Amino-nitrogen/ (g/L)	0.5±0.0	0.5±0.0	0.6±0.0	0.5±0.0	0.5±0.0	0.5±0.0	NS
Ethanol/ (%v/v)	12.6±0.7b	13.0±0.1 a	13.9±0.1a	12.8±0.3	12.8±0.1a	13.2±0.2a	*
Total amino acid/ (mg/L)	3117.7± 88.8d	3537.0± 153.4 b	3684.3± 80.5a	3377.9± 101.1c	3489.0± 106.5b	3499.5± 104.1b	**
Total soluble solids (°Brix)	10.7±0.4 a	9.1±0.3c	8.1±0.5d	10.3±0.5ab	9.9±0.2b	8.7±0.3cd	*

Different lowercase letters in the same row indicate significant differences ($p < 0.05$, LSSD text)

Significant difference at $*P < 0.05$ and $**P < 0.01$, respectively; NS, not significant

Questions on Text

What are the groups of Chinese rice wine?

Words to Watch

nutrition [njuˈtrɪʃən] n. 营养；滋养

beverage [ˈbɛvərɪdʒ] n. 饮料
spontaneous fermentation 自然发酵
koji [ˈkəudʒi] n. 曲；日本酒曲；清酒曲
structure [ˈstrʌktʃə] n. 结构，构造；体系
dominant strain 优势菌株
insufficient [ˌɪnsəˈfɪʃənt] adj. 不充分的；不足的；不够重要的

Phrases and Patterns

more than 超出；比……更加
divided into 分为
result from 由……引起

Supplementary reading

Chinese rice wine, a natural and non-distilled wine, is very popular in China and its market is speedily increasing. The annual consumption is about 1.4 million tons. Hitherto, the Chinese rice wine brewing process is mainly controlled by experienced technician rather than by scientific instruments. This technician control method causes each batch of Chinese rice wine with different flavors. Currently, how to standardize all batches of Chinese rice wine with the same flavor is still an unresolved issue. Good taste becomes more important than ever for the Chinese rice wine. Young drinkers have more choices for drinks. Consequently, the wine should be with good and consistent taste to attract more customers. It is thus very important to study the effects of temperature on Chinese rice wine brewing.

Lesson 2   Jiuqu

Objectives

At the end of this lesson, students should be able to:
1. Briefly introduce what Jiuqu is.
2. Understand the production process of Jiuqu.
3. Understand the microorganisms in Jiuqu.

Text

Jiuqu is a kind of fermentation starter for the production of Chinese sweet rice wine (CS-

RW), which is a traditional Chinese alcohol beverage, possessing a desirable flavor and high nutritional value, including peptides, oligosaccharides, vitamins, amino acids, and organic acids. The starters are the mixtures of yeasts, molds, and bacteria, which produce plenty of enzymes for cellular metabolisms and subsequent small molecule generation, which contribute to final quality of the products.

Over thousands of years, humans have mastered the natural acids and esters fermentation technique of cultivating functional microbiota on different raw materials. The microbiota in Jiuqu can produce large amounts of acids and esters. Manipulation of environmental variables such as nutrient, humidity, temperature, and aeration results in a wide spectrum of fermented acids and esters. In East Asia, especially in China, cereals are commonly used for the production of acids and esters with characteristic flavours. Metabolism of various microorganisms in the acids and esters fermentation improves the nutritional value, sensory properties, and bioactivities of raw materials.

Even though Jiuqu is designed to ferment rice in China, and the microbial composition of Jiuqu is not fully understood yet, the results indicated that it is useful to utilize combinations of microorganisms to optimize the rice wine production.

Questions on Text

Why Jiuqu can produce acids and esters?

Words to Watch

Jiuqu   n. 酒曲
fermentation [ˌfɜːmɛnˈteɪʃən] n. 发酵
traditional [trəˈdɪʃənl] adj. 传统的；惯例的
mold [məʊld] n. 霉菌
bacteria [bækˈtɪərɪə] n. 细菌
microbiota [maɪkrəʊbaˈɪɒtə] n. 小型生物群，微生物区
microorganism [ˌmaɪkrəʊˈɔːgənɪzəm] n. 微生物

Phrases and Patterns

rice wine 米酒
mixtures of 是……的混合
result in 导致
be used for 用于

even though 尽管

be useful to 对……有用

Supplementary reading

Saccharomyces cerevisiae is one of the most commonly used yeast for the fermentation of kiwi juice in China; however, there are a few problems associated with *S. cerevisiae* fermented kiwi wine such as lack of ample fruitiness, high acidity, low typicality, high methanol content. Some researchers believed that *S. cerevisiae* is not the ideal one for producing kiwi wine. Mixed microorganism fermentation may improve the quality of kiwi fruit wine as there are many successful examples on other fruit: reported that compared to pure fermentation, mixed fermentation with *Starmerella bacillaris* and *S. cerevisiae* improved the quality of Barbera wine because of the synergistic effect of the two microorganisms, the glycerine content can be increased to 8.8~9.8 g/L and acetic acid content can be reached up to 0.4 g/L. Garcia found that mixed fermentation with *Debaryomyces vanriji* and *S. cerevisiae* significantly improved the concentration of several acids, alcohol, ester, and terpinol, and therefore better volatile profile of grape wine was detected by mixed fermentation. Minnaar, Jolly, Paulsen, Du Plessis, and Van Der Rijst reported that the titratable acid of Kei-apple juice fermented by Schizo *Saccharomyces pombe* and *S. cerevisiae* yeasts decreased by 70% compared to single culture fermentations. According to these reported researches, kiwi wine produced with mixed fermentation may improve its final quality. Therefore, the aim of this work was to investigate the organic acid, flavor substance, methanol and sensory quality of kiwi wines produced by the mixed fermentation of Jiuqu and *S. cerevisiae*.

China is well-known for its very long history of alcoholic production, traceable to as early as 9000 years ago. Analyses of microfossil remains on pottery vessels indicate that the early Neolithic people used globular jars to make cereal-based alcoholic beverages, and also developed diverse fermentation techniques some 8000 years ago. This development is exemplified by discoveries of a beer made from sprouted millet in Linkou and of another made from Jiuqu or Qu (from moldy grains and grasses) at Guantaoyuan, two sites located along the Wei River valley. During the middle Neolithic period, a new vessel type, jiandiping amphora, became widespread, representing a typical artifact of the Yangshao culture in the Yellow River region, North China. Some amphorae appear to have been used for alcoholic fermentation and storage, or as drinking utensils. Residue analysis of amphorae from three Yangshao sites in the Wei River valley all point to sprouted millet as the beer's main constituent grain.

Lesson 3 Shaoxing Rice Wine

Objectives

At the end of this lesson, students should be able to:
1. Understand the technology of rice wine.
2. Have a good understanding of microorganisms in the fermentation process of rice wine.
3. Have a good understanding of Chinese famous rice wine.

Text

Huangjiu, also named Chinese rice wine, a very special alcoholic beverage in China which dates back to more than 5000 years ago. It is popular in southeast China and Southeast Asia due to its unique aroma. The unique brewing technology handed down through generations produces rice wine with a bright brown color, subtle sweet flavor, and low alcohol content. Chinese rice wine has been honored as the national banquet wine because of its high medicinal value and nutritional benefits. There are many kinds of Chinese rice wine in China according to different geographical origins, such as Shaoxing rice wine, Zhejiang Jiashan wine, Shanghai rice wine and others. Among them, Shaoxing rice wine is the most famous and representative Chinese rice wine with its unique flavor characteristics.

Shaoxing rice wine is fermented from rice with wheat Qu as saccharifying agent and yeast as fermentation starter. The feature of process refers to simultaneous saccharification fermentation. In general, The brewing of traditional Chinese rice wine was carried out using open fermentation and pottery jars. In that case, the saccharification of starch and the fermentation of sugars are conducted at the same time, including two major stages: raw rice pretreatment and alcohol fermentation. The traditional pretreatment is conventionally performed by soaking low-amylose or waxy paddy rice in water at room temperature for 3~4 days for maximum hydration. And the fermentation refer to large unknown microorganisms are intentionally added to the fermentation mash. They build a complex symbiotic relationship during the brewing process and form a special flavor to Chinese rice wine. Among these microorganisms, bacteria play a prominent role in producing the flavor compounds of Chinese rice wine and determined the overall quality of the final product.

It was demonstrated that the fermentation of Chinese rice wine is a multi-strains co-fermentation process associated with more than ten genera. The ten dominating genera had significantly different change trend during the fermentation process. *Bacillus* and *Lactobacillus* were the most dominant bacteria. Totally, 64 volatile compounds were identified during fermented process. The volatile compounds were mainly esters, alcohols, carbonyl compound and phenols.

As a traditional alcoholic beverage, Chinese rice wine with a unique flavor and high nutritional value has been popular in China for thousands of years. With the development of economic globalization, Chinese rice wine has been sold to the international market. Meanwhile, with the improvement of the living level of Chinese people, higher quality of Chinese rice wine was requested.

Words to Watch

aroma [əˈrəʊmə] n. 芳香
flavor [ˈfleɪvə] n. 风味；香料；滋味
alcohol [ˈælkəhɒl] n. 酒精，乙醇
content [ˈkɒntɛnt] n. 内容，目录；满足；容量
geographical [ˌdʒɪəˈɡræfɪkəl] adj. 地理的；地理学的
origin [ˈɒrɪdʒɪn] n. 起源
simultaneous [ˌsɪməlˈteɪnɪəs] adj. 同时的；联立的；同时发生的
bacillus [bəˈsɪləs] n. 杆菌
lactobacillus [ˌlæktəʊbəˈsɪləs] n. 乳酸菌
compound [ˈkɒmpaʊnd] n. 化合物；混合物

Phrases and Patterns

due to 由于
refer to 提到
be honored as 被誉为
associate with 与……联系

Supplementary reading

Shaoxing rice wine is fermented from rice with wheat Qu as saccharifying agent and yeast (*Saccharomyces cerevisiae*) as fermentation starter. The feature of process refers to simultaneous saccharification fermentation. In the beginning of fermentation, large unknown microorganisms are intentionally added to the fermentation mash. They build a complex symbiotic relationship during the brewing process and form a special flavor to Chinese rice wine. Among these microorganisms, bacteria play a prominent role in producing the flavor compounds of Chinese rice wine and determined the overall quality of the final product. The reveal of the bacterial community is a prerequisite for understanding and controlling of Chinese rice wine brewing.

However, there are few reports about bacterial community of Chinese rice wine, and most of them were studied mainly using cultivation-dependent methods such as cell cultures and colony counting. These methods provide a traditional and simple way to study bacterial communities of Chinese rice wine, but isolation media may be only suitable for certain types of microbes, as most of microbes cannot be cultivated under standard laboratory protocols till now. Thus, various molecular methods in different alcoholic beverage have been developed in recent years, such as denaturing gradient gel electrophoresis (DGGE), cloning and the amplification of rRNA genes using PCR and fluorescence internal transcribed spacer-PCR (f-ITS-PCR). The molecular biology techniques provide precise insight into microbial diversity and a rapid, high-resolution description of microbial communities by targeting ribosomal genes. At the same time, the high-throughput technologies such as next-generation sequencing are rapidly changing the way to study microbial communities. Particularly, the Illumina MiSeq pyrosequencing, a simple and rapid method of studying microbial community, can sequence hundreds of thousands of nucleotides at the same time. It is yielding comprehensive information about microbial communities in various environments and provides relatively quantitative comparisons of microbial communities at depths previously unattainable.

Lesson 4　Jimo Lao Jiu Rice Wine

Objectives

At the end of this lesson, students should be able to:
1. Briefly introduce what Jimo Lao Jiu rice wine is.
2. Learn the history of Jimo Lao Jiu rice wine.
3. Understand the brewing process of Jimo Lao Jiu rice wine.

Text

Jimo Lao Jiu has a long history of 4000 years, originated in the Shang Dynasty, prevailed in the Spring and Autumn and Warring States period, has experienced vicissitudes, and continues to this day. It has a really long history, unique craftsmanship, excellent crystal quality, and splendid culture. The brand culture has become the northern representative of China's rice wine, and it can be called a historical wine and a Chinese treasure.

Jimo Lao Jiu Rice wine is made of rice, millet, corn, wheat, etc. as the main raw materials. After soaking, cooking, adding Jiuqu, saccharification, fermentation, pressing, filtering, decoction, storage, blending, it has a special color, aroma and taste and it is one of the three oldest wines in the world. Rice wine is rich in nutrition, not only contains a variety of essential amino acids, but also contains nutrients such as alcohols, acids, lipids, carbonyl compounds and phenols.

In addition, in terms of technical process, Jimo Lao Jiu adopts low-temperature fermentation method to produce dry wine with high alcohol content and low sugar content. The introduction of sub-tank fermentation is used in the summer when the temperature is high, which is beneficial to control the fermentation temperature and prevent mash rancidity. The trough fermentation method not only saves the construction area, but also is more conducive to the control of the fermentation process and ensures the product quality. Jimo Lao Jiu has independent intellectual property rights and 31 national patents. Saccharification technology transformation, dry Jimo Lao Jiu, sea cucumber Jimo Lao Jiu and other technological innovation projects have filled the gaps at home and abroad. The annual R&D of new products has an annual economic benefit of more than 30 million yuan.

Questions on Text

What are the nutritional values of Jimo Lao Jiu?

Words to Watch

craftsmanship [ˈkrɑːftsmənʃɪp] n. 技术；技艺
saccharification [səˌkærɪfɪˈkeɪʃən] n. 糖化（作用）
press [pres] v. 压榨
filter [ˈfɪltə] v. 过滤，滤除
decoction [dɪˈkɒkʃən] n. 煮出法
lipid [ˈlɪpɪd] n. 脂肪，油脂
carbonyl [ˈkɑːbənɪl] n. 羰基
intellectual [ˌɪntɪˈlektjʊəl] adj. 智力的；聪明的；理智的

Phrases and Patterns

originate in 起源于
prevail in 在……盛行

be made of 由……制造

be rich in 富于，富含

not only... but also... 不仅……而且

in addition 此外

in terms of 就……而言；在……方面

be used in 用于

Supplementary reading

Jimo Lao Jiu rice wine, which has a long history, is one of the most ancient beverages in the world. Jimo Lao Jiu rice wine is fermented from rice, millet and other cereals by mixed saccharifying starters, such as wheat starter and distiller's yeast. In the traditional rice wine production industry, glutinous rice is the most widely used raw material because it easily absorbs water and gelatinizes. Accompanying the continuous development of the wine production industry, an increasing number of rice varieties have become available. Compared with glutinous rice, japonica rice has a lower price and a higher yield. Japonica rice could produce rice wine with a higher amino acid content because of its high protein content.

Problems in traditional rice wine production limit its development, such as a long production cycle and a large amount of wastewater generation. In general, the main production process of rice wine includes three steps: soaking and steaming, saccharification and alcohol fermentation. Among these steps, soaking is one of the most important processes and will directly affect the quality of the resulting rice wine. Traditionally, soaking takes 2 or more days, which means a longer production cycle and more energy consumption of the process. Considering the above problems, many researchers have worked to develop a series of new technologies to replace traditional manufacturing techniques. Kuo et al. employed extruding technology instead of traditional soaking and cooking methods and found that extruding could increase alcohol production. Chen and Xu brewed rice wine by roasting rice with superheated steam and found that the gelatinization effect was better and the amino nitrogen content in the wine was lower, which resulted in a lighter flavor.

Vacuum soaking is a new technology that involves the application of a negative pressure on the surface of food so that the external solution can penetrate the food much more quickly than with soaking at atmospheric pressure. This technique can help to improve the product quality, accelerate the soaking process and save energy. Vacuum soaking technology is mostly used in the processing of fruit and vegetable products, showing different effects on the physicochemical properties and microstructures of the products than those of conventional soaking. In recent years, studies on rice indicated that soaking under vacuum conditions could shorten the rice soaking time, enhance water absorption and ameliorate rice properties.

Lesson 5   Sake

Objectives

At the end of this lesson, students should be able to:
1. Briefly introduce what Sake is.
2. Understand the flavor composition of sake.
3. Learn about the history of sake.

Text

Sake production in Japan is regulated under the jurisdiction of the National Tax Agency, not the Ministry of Agriculture, Forestry and Fisheries. Japanese sake is a traditional alcoholic beverage made from rice, koji, and water. Simultaneous saccharification of rice by koji (*Aspergillus oryzae*) and alcohol fermentation by yeast (*Saccharomyces cerevisiae*) lead to the formation of a variety of ingredients of Japanese sake that give it its characteristic taste. Sake contains more than 280 metabolites that affect its quality. The metabolite composition of sake depends on the combination of raw materials and sake-making parameters (e.g., rice species, rice polishing ratio, water quality, koji mold, yeast strains, sake mash fermentation methods) used during production. Thus, investigating the association between sake metabolites and sake-making parameters is important for controlling the taste and quality of sake.

Sake is a traditional Japanese alcoholic drink with a tradition lasting more than 1300 years. At present, more than 1000 sake brewing companies are widely distributed throughout Japan. Although the sake production method is very similar across these companies, the flavor, taste, and quality of the sake differs. Koji, moto (seed mash), rice, and water are used for sake production, and no bacteria are added. Koji is produced using the koji mold *Aspergillus oryzae*, which converts rice starch into sugars. The fermentation starter moto is produced using koji and the sake yeast *Saccharomyces cerevisiae*, which converts sugars to ethanol.

Sake lees (Sake-kasu) are the white paste derived from the filtered residue of Japanese Sake (rice wine) mash. Sake lees are a rich source of proteins, peptides, amino acids, carbohydrates, fiber, fat, ash, and vitamins. Consuming sake lees as toasted cakes and soup has a long history. Nowadays, Sake lees are added to pickled vegetables and fish to impart a fruity aroma and flavor in them. They are also used as flavoring agents for beverages, confectionery and various processed foods. Sake lees contain various enzymes and metabolites derived from molds and yeasts such as *Aspergillus oryzae* and *Saccharomyces cerevisiae* used in Japanese Sake brewing. *Aspergillus oryzae* produces acid proteases and acid carboxypeptidases. *Saccharomyces cerevisiae* generates carboxypeptidases, proteinases, and aminopeptidases. Sake lees and their constituents have been reported to improve hepatic lipid accumulation, reduce blood pressure, mitigate hyperalgesia, prevent allergic rhinitis-like symptoms, and suppress acute alcohol-induced liver injury. Several studies reported reductions in the viable counts of aerobic bacteria in raw ham treated with Sake lees and altered texture in squid meat cured with Sake lees based on sensory tests, texturometry, and protein decomposition analyses. Despite the well-established biotechnological and nutritional advantages of Sake lees, their applications in food processing have seldom been explored.

Questions on Text

What are the main ingredients of Sake?

Words to Watch

jurisdiction [ˌdʒʊərɪsˈdɪkʃən] n. 司法权，审判权，管辖权
characteristic [ˌkærəktəˈrɪstɪk] adj. 典型的；特有的；表示特性的
peptide [ˈpeptaɪd] n. 多肽类；缩氨酸
carboxypeptidase [kɑːbɔksiˈpeptideɪs] n. 羧肽酶
hyperalgesia [ˌhaɪpərælˈdʒɪzɪə] n. 痛觉过敏

Phrases and Patterns

between... and... 在……和……之间
be important for 对……很重要
at present 目前

Supplementary reading

Sake yeast strains possess characteristics that are favorable for sake brewing, i.e., production of high concentration of ethanol and rich aroma. Sake yeast has been identified as *Saccharomyces cerevisiae* and has a heterothallic life cycle. Therefore, cross breeding of sake yeast strains can result in a combination of different features derived from various strains. However, the major obstacle in successful cross breeding of sake yeast is the difficulty in obtaining haploid strains of sake yeast because sake yeast demonstrates extremely poor sporulation under sporulation-inducing conditions.

In the sake industry, parameters such as rice cultivars, yeast species, and fermentation temperature are altered according to the rice polishing ratio. For example, when the rice polishing ratio is low, Yamadanishiki (sake rice) and sake yeast K1801 (suitable for Ginjo sake) are used. The combined effects of these parameters on the sake metabolome have not been examined. In this study, we analyzed several types of commercial sake using our sake metabolome analysis method and revealed that sake metabolites are correlated with the rice polishing ratio. As previously mentioned, some parameters tend to change with the rice polishing ratio. Thus, to examine the statistical correlation between each parameter, we performed combination experiments to evaluate sake-making parameters, such as rice polishing ratio, rice cultivars, and yeast strains used in small-scale fermentation. The results revealed that many metabolites are affected by various parameters and that these parameters have combined effects on sake metabolites.

Chapter 2
Distilled Alcoholic Beverages

Unit 1　Chinese Baijiu
Unit 2　Brandy
Unit 3　Whisky
Unit 4　Other Distilled Alcoholic Beverages

Unit 1 Chinese Baijiu

Lesson 1 Introduction

Objectives

At the end of this lesson, students should:
1. Know what the specific Baijiu is.
2. Familiar with the brewing process of Baijiu.
3. Know the most commonly used enzyme preparations.

Text

All raw materials containing starch and sugar can be brewed liquor, but different flavors of liquor made from different raw materials. Grain sorghum, wheat, corn; sweet potato and cassava; sugar cane and sugar beet residue, waste molasses can be used to make wine (Figure 2-1).

Figure 2-1 Cereals

In addition, sorghum bran, rice bran, bran, rice water, starch residue, sweet potato crumbs, beetroot tails etc. can be used as alternative raw materials. Wild plants such as oak seed, *Jerusalem artichoke*, *Pyrus betulaefolia*, Jin Yingzi, etc. can also be used as a substitute raw material. The traditional Baijiu brewing process in country is solid-state fermentation method, some auxiliary materials need to be added during fermentation to adjust starch (Figure 2-2). Concentration maintain the softness of the mash, and the slurry water. Commonly used the auxiliary materials are rice husk, grain bran, corn cob, sorghum husk, flower raw hides etc. In liquor, starch needs to be hydrolyzed by a variety of amylases. Using produce sugar that can be fermented, so that it can be fermented used by distiller's yeast. This process is called saccharification, the sugar used the chemical agent is called distiller's yeast (or Baijiu distiller's yeast, saccharified distiller's yeast). Distiller's yeast contains starch. The main raw material is used as a medium to cultivate a variety of molds and accumulate.

A large amount of amylase is a crude enzyme preparation. Currently the countrified distiller's yeast includes Daqu (for the production of famous and high-quality Baijiu) (Figure 2-3A), Xiaoqu (for the production of Xiaoqu Baijiu) and bran qu (for the production of bran qu Baijiu). The most widely used in production is bran qu. In addition, sugar is yeast. The alcoholic enzyme secreted by the bacteria transforms into alcohol and other substances, it's called alcohol fermentation, the fermentation agent used in this process called distiller's yeast. It is based on sugar-containing substances as the culture medium; yeast has undergone fairly pure expansion culture, and the resulting ferment. Large-tank distiller's yeast is used in production. There are two types of Baijiu production in China: solid fermentation and liquid fermentation. In solid-state fermentation of Daqu, Xiaoqu, Bran qu and other processes, the Bran qu occupies a large proportion in the products.

Figure 2-2 Solid state fermentation

Figure 2-3A Ferment Agent

Questions on Text

What are brewing ingredients?

Words to Watch

sorghum [ˈsɔːgəm] n. 高粱；高粱米
cassava [kəˈsɑːvə] n. 木薯
molasses [məˈlæsɪz] n. 糖蜜；糖浆
fermentation [ˌfɜːrmenˈteɪʃn] n. 发酵
auxiliary [ɔːgˈzɪliəri] adj. 辅助的；备用的
slurry [ˈslʌri] n. 泥浆
hydrolyze [ˈhaɪdrəlaɪz] v. 水解
amylase [ˈæmɪleɪz] n. 淀粉酶
saccharification [sækərɪfɪˈkeɪʃən] n. 糖化；糖化作用
cultivate [ˈkʌltɪveɪt] v. 培养
enzyme [ˈenzaɪm] n. 酶
secrete [sɪˈkriːt] v. 分泌；隐藏

Phrases and Patterns

be used as 用作
a variety of 各种各样的
be based on 基于

Supplementary Reading

The spirits culture

In the long history of China, Chinese spirits, just like the ancient Chinese culture, has deeply influenced people's life and culture in Chinese daily life, for example, the traditional Baijiu brewing. The history of spirits can be dated back to the old times. In historical records there were some records about spirits: "King Zhou made spirits pool and hung meat in a forest; drinking all night long." In the book of songs there were some poems about spirits: "October is the time to harvest the grain to make spirits, and to use the spirits to celebrate the longevity of the senior." Those records indicated that spirits had a history at least five thousand years.

Spirits brewing

It has a history of thousands years and the traditional fermenting technique developed gradually into a mature level. Even in nowadays, the natural fermenting technique has not disappeared completely and something about such technique still remains its mystery. People made spirits based on their experience. Everything was done by manual labor and has no reliable standards to assure the quality of spirits.

Spirits testing

People use their visual, olfactory, taste and touch to evaluate the quality of spirits, distinguish good or bad spirits, to tell the features in it. That's what we called spirits tasting. Judging the quality of spirits according to some standards of physical and chemical analyses is not enough. Till now, no equipment can tell all flavors in a comprehensive and accurate way. Whether the color, flavor and taste of a certain spirits can be accepted by the people of a certain country or region should be decided by spirits tasting itself. Spirits tasting is a science. It is also a traditional craft handed down by our ancestors.

Function of spirits

Spirits was used to celebrate different festivals, wedding ceremonies and birthday parties, to memorize the departed, welcome and send off relatives and friends, congratulate the good news and to get rid of anxiety, cure diseases and prolong life, both for the king and ordinary people.

Three cups of spirits will make the spirits lovers very comfortable and happy, feeling like the immortal. All unhappiness will disappear. At that time, people's nature will return. The evil will become kind; the cowards will be brave; the stupid will become clever. The poor can express their anger; the strong can boost their spirits and the powerful can make their decisions.

Upper limit of drinking spirits

From the standpoint of healthy life, one should drink spirits properly, therefore Baijiu is the best chose. There is no standard how much one can drink mostly. Generally speaking, a normal person can drink 0.6 ~ 0.8 milliliters pure alcohol each time for every 1 kilogram of his body weight, such amount will not only do no harms, but also make a person relaxed and happy. So at this scale, a 60kg person can drink:

Beer: about 1.5 to 2 bottles;

Wine (12 degrees): 300~400mL;

Rice wine: 225~300mL;

Brandy, whisky: 80~110mL; Baijiu (65 degrees) 50~70mL.

The future

The first is culture:

Guoguanjiu contains Chinese culture and redefines Chinese Baijiu. As we all know, Chinese Baijiu is closely connected with Chinese culture. Baijiu has always represented the culture of the elite. There are many cultures about Chinese liquor, but in fact most of them are the product culture or fabricated culture of enterprises, which is far from the culture of the elite. Guoguanjiu uses the resources of Chinese artists and cultural elites to develop liquor products, so that art and Baijiu are perfectly integrated.

The second is quality:

China Yuanjiu Commune, only making Chinese grain and qualified Baijiu. After years of development, Chinese baijiu needs the ultimate craftsmanship so that consumers can obtain better products and better experience, so that they can achieve their status in the market. At the same time, as a symbol of national culture, it has been defeated by various quality crises. Therefore, in this confused environment, the emergence of original wine can make consumers re-recognize Chinese Baijiu. Baijiu is between daily necessities and optional consumer goods, and has its own unique consumption scenarios. Such as, banquets, gifts, collection and self-drinking, etc. They are all long-term needs.

The overall sales volume of the Baijiu industry is gradually declining, but due to two factors, the price increase of the Baijiu industry and the high-end consumption of Baijiu, the overall revenue of the Baijiu industry is still rising steadily, and the net profit is increasing year by year. Among them, high-end Baijiu benefited from the two factors of high-end consumption structure and price increase, achieving both volume and price increases. In the future, the overall profit of the Baijiu industry will mainly come from price increases, and mid-to-high-end Baijiu will continue to benefit from the high-end consumption structure of Baijiu.

Vocabulary

amylose [ˈæmɪləʊs] n. 直链淀粉
amylopectin [ˌæmələʊˈpektɪn] n. 支链淀粉
glycerin [ˈɡlɪsərɪn] n. 甘油（丙三醇）
hemicellulose [ˌhemɪˈseljʊləʊs] n. 半纤维素
lignin [ˈlɪɡnɪn] n. 木质素
maltose [ˈmɔltəʊz] n. 麦芽糖
dextrin [ˈdekstrɪn] n. 糊精
oligosaccharide [ˌɔlɪɡəʊˈsækəraɪd] n. 低聚糖
tartaric acid n. 酒石酸
malic acid [ˈmeɪlɪkˈæsɪd] n. 苹果酸

succinic acid   n. 琥珀酸
lactic acid   n. 乳酸
higher fatty acid 高级脂肪酸
guaiacol ['gwaɪəkɒl] n. 愈创木酚
polyalcohol [pɒlɪ'ælkəhɒl] n. 多元醇
pit mud 窖泥
polyphenol [pɔli'fi:nɔl] n. 多酚
precursor substance 前体物质
arginine ['ɑdʒəˌnin] n. 精氨酸
citrulline [sɪ'trʌlin] n. 瓜氨酸
urethane ['jʊrəθeɪn] n. 氨基甲酸乙酯
pyrazine compound 吡嗪类化合物
carbon to nitrogen ratio 碳氮比
rhizopus ['raɪzəʊpəs] n. 根霉
ester ['estə(r)] n. 酯类物质
ethyl acetate 乙酸乙酯
ethyl lactate 乳酸乙酯
ethyl butyrate 丁酸乙酯
ethyl hexanoate 己酸乙酯
aromatic substance 芳香物质
isoamyl alcohol   异戊醇
furfural ['fɜfəræl] n. 糠醛
aromatic compound 芳香族化合物
corn [kɔ:n] n. 玉米
wheat [wi:t] n. 小麦
barley ['bɑ:li] n. 大麦
pea [pi:] n. 豌豆
excipient [ɪk'sɪpɪənt] n. 赋形剂
rice husk   n. 稻壳，谷壳
sorghum shell   n. 高粱壳
hemicellulase [hemɪ'seljʊleɪs] n. 半纤维素酶
polyphenol oxidase 多酚氧化酶
iron-sulfur protein 铁硫蛋白
amylase ['æmɪleɪz] n. 淀粉酶

saccharifying enzyme 糖化酶
xylanase [ˈzaɪləneɪs] n. 木聚糖酶
amino acid dehydrogenase 氨基酸脱氢酶
alcohol fermentation 酒精发酵
saccharify [səˈkærɪfaɪ] vt. 糖化
gelatinize [dʒɪˈlætɪnaɪz] v. （使）成胶状
liquefaction [ˌlɪkwɪˈfækʃən] n. 液化
rectification [ˌrektɪfɪˈkeɪʃn] n. 精馏
solid state fermentation 固态发酵
semi-solid fermentation 半固态发酵
double fermentation 双重发酵
liquid fermentation 液态发酵
catabolic pathway 分解代谢途径
proton motive force 质子动力势
siderophore [ˈsɪdərəfɔː] n. 铁载体
hydroxamate [haɪdˈrɒksæmət] n. 异羟肟酸
Streptococcus [ˌstrɛptəʊˈkɒkəs] n. 链球菌属
Lactobacillus [ˌlæktəʊbəˈsɪləs] n. 乳杆菌属
Leuconostoc [ˌljuːkəˈnɒstɒk] n. 明串珠菌属
casein [ˈkeɪsiːɪn] n. 酪蛋白
glycolysis [glaɪˈkɒlɪsɪs] n. 糖酵解
anaerobic oxidation 无氧氧化
dehydrogenase [diːˈhaɪdrədʒəneɪs] n. 脱氢酶
cyanide [ˈsaɪənaɪd] n. 氰化物

Lesson 2　Daqu

Objectives

At the end of this lesson students should：

1. Briefly know the preparation process of Jiang-flavor Daqu.
2. Know raw material for preparing Jiang-flavor Daqu.
3. Know how to choose Muqu of flavor Baijiu.

Text

Daqu contains Nong-flavour Daqu, Mild-flavour Daqu, Esterified Daqu and Jiang-flavour Daqu, etc (Figure 2-3B, 3C, 3D, 3E). The preparation process of Jiang-flavour Daqu is high-temperature koji making, using pure wheat as raw material, through artificial or mechanical forming, cultivating, turning, storage and other stages to convert it into a collection of microbial flora, multiple enzymes and volatilization. An intermediate Baijiu making product that combines components and precursors. After the raw material wheat is dusted and impurity removed, adding 2% ~ 3% water to stir and moisten it for 3 to 4 hours. After the moistening is completed, using a fresh mill to crush the wheat into 'plum petal' flakes. Add the Muqu and water to the raw materials, and stir the koji. The Muqu selects the high-quality Daqu from the previous year, and the amount of Muqu is 4% ~5% in summer and 5% ~8% in winter. The amount of water added is generally 37% ~40% of the raw material. After the raw materials are stepped into the koji, the koji is put into the room for accumulation culture. When mildew grows on the surface of the curved billet, perform the first turning. After the first turning, perform the second turning after 7~8t. When the curved billet is accumulated and cultivated in the curved chamber for about 25t, the moisture is discharged. When it grows to about 40t, it will be the finished Jiang-flavor Daqu. After the raw wheat is moistened and crushed, the Muqu is added to mix the raw materials. The Muqu is a high-quality Daqu from the previous year, and the Muqu is required to be crushed as finely as possible. The addition amount of summer Muqu is 4%~5%, and the addition amount of winter Muqu is 5% ~8%. Generally, the amount of water added when mixing the koji material is 37%~40% of the raw material. However, adding too much or too little water will affect the quality of the koji. When the amount of water is too much, the slab is easy to be pressed too hard, the mold grows vigorously, the temperature rises quickly and violently, the temperature is not easy to lose, and the water is not easy to volatilize, which affects the fermentation and culture of the house. When the amount of water added is too little, the koji material absorbs water slowly and the koji blank is easy to disperse. Due to the inability to provide the water needed for the production, reproduction and metabolism of microorganisms, molds, bacteria and yeasts cannot grow and reproduce normally, the fermented koji dough is not thorough, and the quality of Daqu is not good. When the wheat is stepped into a slab after being evenly stirred, the pressed slab should be placed for 2~3 hours first, which often called "sweat removal". The surface of the slab is slightly dry and can be placed in the slab for accumulation after hardening to cultivate. The accumulation and cultivation of koji is also strictly required in the production and fermentation process of Jiang-flavor Daqu. The production of process of Jiang-flavor Daqu must be reasonably "stacking" the koji. Before the curved billet enters the curved room, a layer of straw (about 15cm) should be laid on the wall and the ground. The straw is mainly used for heat preservation. The curved billets should be stacked 3 horizontally and 3 vertically, arranged alternately. The distance between the curved billets is generally 1.5~2cm in winter and 2~3cm in summer, separated by straw. The thickness of straw is about 7cm between the curved blank layer and the layer. The horizontal and vertical arrangements of the upper and lower curved blanks should also be staggered to facilitate air circulation and heat preservation, and promote the growth of mold. The pile height is generally 4~5 layers. After arranging one row of billets, arrange the second row of billets next to each other, and finally set aside one row for reversal. After the curd is piled up, cover the top and the surroundings of the curd with straw, and after sprinkling water on the straw, close the doors and windows of the fermentation room, and keep warm and moisturized. Microorganisms are multiplying in the Daqu, and the product temperature is rising. When the temperature of the slab is about 63°C, mildew grows on the surface, and the first turning is possible. After 7~8 days, perform the second reversion. Turning requires the upper and lower, inner and outer layers to be reversed (Figure 2-4).

Figure 2-3B Nong-flavour Daqu

Figure 2-3C Mild-flavour Daqu

Figure 2-3D Esterified Daqu

Figure 2-3E Esterified Daqu

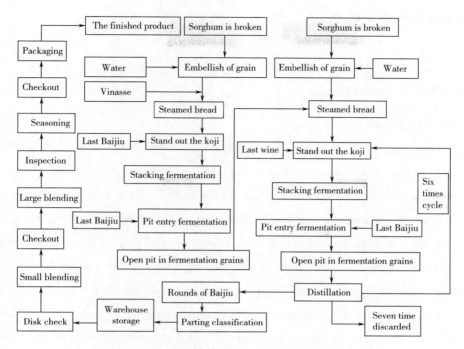

Figure 2-4　The process of Jiang-flavor Daqu flow chart

Questions on Text

Briefly introduce how Daqu is made.

Words to Watch

volatilization ［ˌvɒlətɪlɪˈzeɪʃən］ n. 挥发；发散
precursor ［prɪˈkɜːsə(r)］ n. 前体
moisture ［ˈmɔɪstʃə(r)］ n. 潮气；水汽；水分
vigorously ［ˈvɪɡərəsli］ adv. 精神旺盛地，活泼地
metabolism ［məˈtæbəlɪzəm］ n. 新陈代谢
microorganism ［ˌmaɪkrəʊˈɔːɡənɪzəm］ n. 微生物；微生物群；微生物界
vertical ［ˈvɜːtɪkl］ adj. 竖的；垂直的；直立的；纵向的
staggered ［ˈstæɡəd］ adj. 交错的；错开的
circulation ［ˌsɜːkjəˈleɪʃn］ n. 循环
moisturize ［ˈmɔɪstʃəraɪz］ v. 滋润

Phrases and Patterns

due to 由于

Supplementary reading

The brief introduction of Koji

Our country has a long history of making koji and making Baijiu. In the 'Qi Min Yao Shu' written by Jia Sixie, an outstanding agronomist in the Northern Wei Dynasty, there is an incisive discussion and summary on the production technology of koji making at that time. Describe the brewing raw material formula, koji making process (Figure 2-5), koji specifications, hygienic conditions, heat preservation measures, finished product identification and preservation methods. Very detailed, and there are many discussions on brewing methods. It can be seen from this that Baijiu entered people's lives 5000 years ago, and it has developed to a considerable scale during the Han Dynasty. It can be said that the use of koji to make Baijiu is a great creation of our working people. Our country is the first country to use microbial koji making and distillation technology to make distilled liquor. China's unique koji brewing technique has been well received at home and abroad. According to historical records, Japan, North Korea, Vietnam and other neighboring countries' experience of using koji to make Baijiu was imported from our country. Liquor koji is roughly divided into five categories: wheat koji, which is mainly used for the brewing of rice wine; Xiaoqu, which is mainly used for the brewing of rice wine and Xiaoqu liquor; red koji, which is mainly used for the brewing of red koji Baijiu; Daqu, which is used for distillation brewing of Baijiu. Bran qu, now developed, uses pure strains to inoculate a culture using bran as a raw material (Figure 2-6).

Figure 2-5 Ancient Baijiu making

Brief introduction to the process of Mild-flavour style (Figure 2-3C)

China's three major fragrant liquors: Nong-flavor, Mild-flavor and Jiang-flavor. Among them, Mild-flavor is represented by Fenjiu produced in Xinghua Village, Fenyang City, Shanxi Province, also known as Fenxiang. It uses grains as raw materials, through the traditional solid-state bilateral method, fermentation, solid-state distillation, aging, and blending to produce Baijiu with a compound aroma based on ethyl acetate. The sensory characteristics of this type of wine are: mild aroma, harmonious, mellow and clean. The characteristics of brewing are "Using underground tanks for fermentation, steaming and clearing twice."

Figure 2-6 Bran qu

The brewing technology experience of fragrant Daqu liquor can be summed up into those words: medium—temperature koji making, pottery fermentation, and steaming.

Firstly, Daqu production: After the raw materials (barley, peas, etc.) are appropriately pulverized, the three types of Daqu, namely stubble, Hongqu and post-fire are prepared under the condition that the maximum temperature does not exceed 50℃, and then the three types of Daqu are prepared according to a fixed ratio (3∶3∶4) use after mixing.

Secondly Raw materials are soaked and boiled, using the process of 'steaming twice and clearing'. The whole production process highlights the word 'clear', that is, "one clear to the end", and auxiliary materials must be steamed before use. Fenjiu uses hot water to moisten grits at high temperature and then cooks. The purpose of cooking is to make the starch particles in the raw material further absorb water and expand on the basis of high temperature moisturizing materials, and finally gelatinize completely, which helps the action of amylase; at the same time, it can kill harmful bacteria in raw materials and auxiliary materials, and send out volatile impurities.

Thirdly, bilateral fermentation, saccharification and fermentation proceed simultaneously. In the whole process, the sugar used by the yeast is gradually produced and supplied. The temperature of the fermented material into the tank is 15~21℃, which will weaken the enzyme damage and facilitate the control of the fermentation process. In such a relatively low temperature environment, the action of glucoamylase is slow, and the accumulation of sugar is less, so the growth and metabolism of microorganisms is weak, so that the yeast will not be prematurely placed in a dense metabolic environment, will not be weak and dying, and the overall fermentation efficiency will be high. In addition, because sorghum granules are tightly organized and difficult to saccharify, solid-state fermentation is used, and starch is not easy to be fully utilized. Therefore, the distilled mash is continuously fermented to utilize its residual starch.

Fourthly, Solid-state distillation. The mash from the vat is evenly. Sprinkled into the retort barrel by manual operation to distill out the Baijiu. This simple solid-state distillation method is not only the concentration and separation of alcohol, but also the process of extracting and recombining aroma components. After obtaining 'rice Baijiu' and 'two rice Baijiu', the two liquors were stored for more than one year, and then blended into finished Baijiu.

The future

China is the first Baijiu making country in the world, using koji is the characteristics of Chinese Baijiu, unique in the history of world Baijiu making, is a great invention of China's pioneers. Mr. Ichiro Makiro, a well-known Japanese liquor making expert, once commented: 'China invented Baijiu making, the impact of which is comparable to China's four major inventions.' Koji is china's earliest and most primitive enzymic preparations, and the application of Baijiu is the most basic biological engineering. It plays an important role in the rise of microbial industry in China. Chinese Baijiu has a long history, unique skills, diverse varieties, each has its own autumn. We should use modern scientific theory and technology to better sum it up, inherit it. In order to improve the quality and variety of Baijiu, and save food to reduce consumption, promote the development of China's Baijiu industry efforts.

The ancient Chinese fermentation technology left us valuable experience including unique and mysterious Baijiu technology: the use of raw materials, environment, air, water and other microorganisms, cleverly control its temperature, humidity and time, culture Baijiu multi-microorganisms to reproduce. For example, Shaoxing wheat qu, Maotai etc. The quality is closely related to the quality of Baijiu. They are not only suitable for brewing microbial reproduction, but also provide the necessary ingredients for the production and formation of aromatic components of Baijiu.

Since the 1930s, the traditional craft of big, small, wheat and red koji in China has been basically stereotyped. Wheat koji is mainly used in Huangjiu brewing, such as Shaoxing wheat koji, etc. Its wheat raw materials are the same as Daqu, but there are differences in technology. Now some people call the Daqu the wheat koji, which is confused. Therefore in the Baijiu making, the Daqu and the wheat Qu is reasonable. In scientific research, some scientists such as Wei Fang yu, Sun Xuefu, Fang Xinfang, Jin Peisong, etc. as well as Dalian Institute of Science, Huanghai Chemical Industry Society, etc. began to engage in the summary of traditional processes, Baijiu microbial research and reform, have achieved encouraging results. Since the founding of the New China, with rapid development, especially after the 1980s, the achievements are particularly remarkable, mainly in the following aspects: firstly, the summary of traditional technology improved. Secondly, the application of pure fermentation technology in modern winemaking industry (the application of asphalt and yeast in solid liquor and liquid liquor production, the application of asphalt and yeast in the production of Huangjiu, the application of root mold and yeast in the production of Huangjiu, enzyme preparation and active dry yeast in Baijiu production, etc.). Thirdly, Baijiu making mechanization, automation has greatly developed, Huang Jiu wheat and Hongqu production, and the application of microcomputers and other automation equipment, greatly reduce the intensity of heavy physical labor, improve quality, reduce costs. Most of the production of large koji using the composition mechanism of the blank, the study of successful shelf koji, used in production (Figure 2-7).

Figure 2-7 Baijiu-making mechanization

Although Chinese Baijiu's production has made achievements, it is not yet suitable with the requirements of the development of the Baijiu industry and has a big gap with other domestic industries as the level of advanced technology, so we must work harder in the new century. Firstly the application of modern new technology and new equipment, improve the production of Baijiu production mechanization, automation level, the liberation of heavy manual labor to improve the traditional manual operation and management methods. Secondly, the development of enzyme preparations and active dry yeast and other new varieties of glycation fermentation agents, to further promote its advanced experience, replace and reduce the amount of Baijiu. Explore new processes for all-enzyme brewing. Further, the use of modern biological new technology, the selection of excellent Baijiu making bacteria species, take the path of pure-bred multi-bacterial mixed fermentation, to achieve both rich Baijiu and raw fragrance purposes. For changing the fermentation of a single strain, the fragrance is insufficient phenomenon. We should develop new koji and new technology, improve the content of science and technology, and create new varieties of health Baijiu. Finally, we adopt modern science and technology and theory, summarize and study the traditional koji making process and basic principles, in order to reform the koji, realize the modernization of production, to provide a scientific basis. At the same time, the relevant departments of the state should organize the financial and human resources of factories, scientific research and universities, and study and solve the key technical projects of koji. The Chinese Society for Microbiology, the Food Science and Technology Association, etc. should organize a national symposium on koji, exchange experiences and formulate long-term development plans.

Noun explanation

低温曲 (Low temperature koji): In the process of making koji, the highest temperature is less than 50℃.

曲母 (Qumu): When making Wuliangye Daqu, inoculate a small amount of koji seeds.

上霉 (Mildew): In the process of koji making, plaque (mold mycelium) grows on the surface of the koji.

堆积发酵 (Pile fermentation): In the process of making Jiang-flavor Baijiu, the grains are spread and dried, mixed with koji, and piled up into piles for opening fermentation.

翻仓 (Turn over): During the process of making Wuliangye, adjust the position of the blanks to make them evenly cultivated.

干曲仓 (Dry starter storehouse): Special room for storing koji.

磨曲机 (Grinding machine): In Wuliangye brewing, a machine that uses dry koji to grind.

磨曲 (Grinding): The operation of grinding koji with a koji grinding machine as required.

陈香曲 (Chen Xiangqu): High temperature koji made in August.

扬冷 (Cooling): The operation of rapidly cooling the retort material to the lower temperature.

月桂曲 (Yuegui Qu): Medium and high temperature koji made in December.

八次加曲发酵 (Eight times koji fermentation): Yibin Baijiu (Jiang-flavor Baijiu) has a unique traditional solid-state fermentation process, which means that it takes eight times to add koji.

菊花心曲 (Chrysanthemum Heart Qu): The section of Daqu finished product is chrysanthemum-shaped, commonly known as chrysanthemum heart koji.

八次摊晾加曲 (Eight times to cool and add Qu): A production process of Jiang-flavor

Baijiu, which refers to eight times of drying and adding koji.

春秋曲 (Spring and Autumn Qu): Medium and high temperature koji made in spring and autumn.

混蒸续糟六甑工艺 (Mixed steaming and six retort process): A process technology for the brewing of Yanghe Daqu, which refers to mixing steam and distilling 6 times with the distiller's grains before.

立体制曲 (Three dimensional koji making): The method of producing Daqu by combining the ground and the shelf to cultivate bacteria.

本窖循环 (The cellar cycle): In the brewing process of Baijiu (such as Wuliangye Baijiu), a technique for adding koji.

挥干 (Dry): A phenomenon of drying the koji.

打拢 (Heap): The process of turning the koji over and maintaining a certain temperature.

糟心 (Zao Xin): A phenomenon of drying the koji.

开窖取醅 (Open the cellar to get the mash): The process of turning the koji over and maintaining a certain temperature.

见汽压醅 (See atmospheric pressure): The one-step requirement for the distillation of the upper retort of Daqu Jiang-flavor Baijiu, means that when you see the gas, you must pay attention to the rise of the glutinous rice and the pressure.

曲模 (Qu mold): Mold for forming Daqu.

鼓曲 (Guqu): A kind of koji.

曲筛 (Koji sieve): The curved screen surface on the fixed wedge-shaped rod which is used to remove the fiber or germ in the suspension.

真菌毒素 (Mycotoxin): Secondary toxic metabolites produced by toxin-producing fungi during growth and reproduction.

2-哌啶酮 (2-piperidone): Volatile flavor components produced by microbial fermentation in Jiang-flavor Daqu.

3-羟基吡啶 (3-hydroxypyridine): Volatile flavor components produced by microbial fermentation in Jiang-flavor Daqu.

呋喃甲酸 (Furoic acid): Volatile flavor components produced by microbial fermentation in Jiang-flavor Daqu.

大火曲 (High temperature koji): High temperature koji.

双轮底酿酒发酵工艺 (Double-wheel bottom brewing fermentation process): That is, after the first round of mash fermentation is mature, the scientific method of remixing is used to inject special high-quality koji and microbial strains to improve product quality for the second round of fermentation.

拌料 (Mixing): The process of brewing Wuliangye, mixing five kinds of grains and koji evenly.

陈曲 (Chen Qu): The newly made koji embryo, which is cultivated and stored for a certain period of time under certain temperature and humidity environmental conditions.

出室 (Out of the room): One of Wuliangye's Koji making process.

堆烧 (Pile burning): One of Wuliangye's Koji making process.

翻曲 (Turn over): One of Wuliangye's Koji Making process.

后缓落 (Fall slowly): Temperature control requirements of Wuliangye manual koji making.

糠糟拌料 (Bran mix): Mix Koji with Stuffed Grains.

地面堆积式自然发酵法 (Ground accumulation type natural fermentation method) Wuliangye's traditional "Baobaoqu" koji making process.

架式立体制曲法 (Shelf-style three-dimensional Qu making): Wuliangye completed and put into production in June 1991, the computer-controlled self-control monitor replaced the koji-making process of "artificial ground stacking and natural fermentation koji making method".

潮气 (moisture): The moisture in the room.

倒伏 (Fall down): The phenomenon of overlapped upside down during the cultivation process.

后火曲 (Houhuo Qu): A variety of Fenjiu Daqu, the early fermentation is slow.

红心曲 (Hongxin Qu): A variety of Fenjiu Daqu, the section is blue-white in appearance and red in the middle.

晾曲 (Cool down): Cool down.

清茬清口 (Clear stubble & mouth): A manifestation of Qingcha QU.

清茬曲 (Qingcha Qu): A variety of Fenjiu Daqu, with a smooth appearance, bluish white and slightly yellow in section, and famous for its fragrance and without peculiar smell.

清蒸清烧一排清法 (Steamed and roasted in a row): A method of producing bran yeast Baijiu.

清蒸混入四大甑法 (Four methods of mixing steamed retort): A method of producing bran yeast Baijiu.

清蒸二次蒸法 (Double steaming method): Method for producing Daqu Baijiu.

清烧法 (Burning method): Method for producing Daqu Baijiu.

4-二甲基氨基吡啶（4-Dimethylaminopyridine）：Volatile flavor components produced by microbial fermentation in Nong-flavor Daqu.

3-庚酮（3-heptanone）：Volatile flavor components produced by microbial fermentation in Nong-flavor Daqu.

双戊烯（Dipentene）：Traditional Jiang-flavor type Daqu microorganisms produce aroma components.

1-辛烯-3-醇（1-octene-3-ol）：The volatile components in rice fragrant koji of Baijiu.

羟基丙酮（Hydroxyacetone）：The aroma substances produced by microbial fermentation in Jiang-flavor Daqu.

2-环戊烯酮（2-cyclopentenone）：The aroma substances produced by microbial fermentation in Nong-flavor Daqu.

乙酰胺（Acetamide）：The aroma substances produced by microbial fermentation in Nong-flavor Daqu.

2,6-二甲氧基苯酚（2,6-Dimethoxyphenol）：The aroma substances produced by microbial fermentation in Nong-flavor Daqu, have sweet, woody, medicinal, and smoky aromas.

苯并呋喃（Benzofuran）：The aroma substances produced by microbial fermentation in Nong-flavor Daqu, have an aromatic taste.

吲哚（Indole）：The aroma substances produced by microbial fermentation in Nong-flavor Daqu.

3-甲基吲哚（3-methylindole）：The aroma substances produced by microbial fermentation in Nong-flavor Daqu.

异亮氨酸（Isoleucine）：An amino acid in koji.

乙酸辛酯（Octyl acetate）：The volatile components in rice fragrant koji of Baijiu.

乙偶姻（Acetoin）：The volatile components in rice fragrant koji of Baijiu.

中温制曲（Medium temperature method）：The highest temperature of the koji is not higher than 58℃.

正戊醛（n-valeraldehyde）：The volatile components in rice fragrant koji of Baijiu.

己酸乙酯（Ethyl caproate）：Volatile components in Baijiu koji.

正丙醇（n-propanol）：The volatile components in rice fragrant koji of Baijiu.

苯乙酮（Acetophenone）：Volatile flavor components produced by microbial fermentation in Nong-flavor Daqu.

2-甲基丁醇（2-methylbutanol）：The aroma substances produced by microbial fermentation in Nong-flavor Daqu, has an unpleasant smell.

乙酰基呋喃（Acetyl furan）：The aroma substances produced by microbial fermentation in Nong-flavor Daqu, has sweet, almond, nut, roasted and smoky aromas.

丁内酯（Butyrolactone）：The aroma substances produced by microbial fermentation in Luzhou-flavor Daqu has sweet, almond, nut, roast, smoky, and ester aromas.

3-吡啶醇（3-pyridinol）：The aroma substances produced by microbial fermentation in Luzhou-flavor Daqu have aroma substances, special smell.

二甲氧基苯酚（Dimethoxyphenol）：The aroma substances produced by microbial fermentation in Luzhou-flavor Daqu.

藜芦醛（Veratraldehyde）：The aroma substances produced by microbial fermentation in Nong-flavor Daqu.

呋喃甲醇（Furan methanol）：Volatile flavor components produced by microbial fermentation in Nong-flavor Daqu.

十六酸甲酯（Methyl palmitate）：Volatile flavor components produced by microbial fermentation in Nong-flavor Daqu.

3-甲基吡啶（3-methylpyridine）：The aroma substances produced by microbial fermentation in Nong-flavor Daqu, has an unpleasant smell.

混蒸续糟法（Mixed steaming）：The process used in the production of Nong-flavored koji, the new grain mixed with mash, steamed and fermented.

2-甲基-1-丙醇（2-methyl-1-propanol）：Volatile flavor components produced by microbial fermentation in Nong-flavor Daqu.

丁酸甲酯（Methyl butyrate）：Volatile flavor components produced by microbial fermentation in Nong-flavor Daqu.

2,6-二甲基吡嗪（2,6-Dimethylpyrazine）：Traditional Nong-flavor type Daqu microorganisms produce aroma components.

2,3-二甲基吡嗪（2,3-Dimethylpyrazine）：Volatile flavor components produced by microbial fermentation in Nong-flavor Daqu.

续糟法（Xu Cha method）：Daqu Baijiu Classification Method.

清糟法（Qing Cha method）：Daqu Baijiu Classification Method.

清糟加续糟法（Qing Cha & Xu Cha method）：Daqu Baijiu Classification Method.

踏曲（Ta Qu）：It is the advancement and extension of straw package koji.

固态配醅（Solid mash）：Fermentation method used in Sichuan Xiaoqu Baijiu.

混曲发酵（Mixed Koji fermentation）：Fermentation in which several kinds of koji are mixed and added to the fermented grains in the production of liquor.

去氧胆酸钠（Sodium Deoxycholate）：Mainly used as a selective inhibitor of bacterial culture medium, often used in Daqu analysis experiments.

二甲基亚砜（Dimethyl sulfoxide）：A sulfur-containing organic compound that is soluble in most organic substances such as ethanol, propanol, benzene and chloroform. It is known as the "universal solvent" and is often used in Daqu analysis experiments.

两段冷凝工艺（Two-stage condensation process）：During the distillation of Daqu Baijiu, the first stage controls the condensation temperature of 60~70℃, and the second stage of cooling temperature is 20℃, which can ensure the steaming of beneficial flavor and increase the amount of impurities.

大小曲法（Mix big and small Qu）：That is, the mixed koji method, such as Dongjiu is Xiaoqu Baijiu, Daqu fragrance, and string fragrance distillation.

复式串香法（Double incense method）：That is, after the Baijiu is brewed according to the method of Xiaoqu Baijiu, it is made into a bottom pot of water and steamed with Daqu method.

双醅串香法（Double mashed incense method）：That is, put the fermented mash of Xiaoqu into the lower layer of the Baijiu bottle, and cover it with the mash made by Daqu method for distillation.

小曲酿造（Xiaoqu Brewing）：Liquor Xiaoqu Brewing.

大曲（Daqu）：Using wheat or barley pea as raw materials, crushed and pressed into brick-like blanks, put into the curved room, enriched with various microorganisms in nature, in the room mature at a certain temperature and humidity.

麸曲（Bran koji）：It adopts pure mold strains and uses bran as raw material to be cultivated through artificial temperature control. It mainly plays a role of saccharification in the Baijiu production process.

小曲（Xiaoqu）：Also known as Jiuyao, Baiyao, Jiubing, it is made of rice noodles (rice bran) as raw materials, added with a small amount of Chinese medicine called polygonum vulgare, inoculated with Qumu, and artificially controlled culture temperature. The shape is smaller than Daqu.

三高三长（Three high and three long）："Three highs" refers to high-temperature koji making, high-temperature accumulation and fermentation, high-temperature flowing Baijiu. "Three-length" means that Daqu is stored for no less than six months, the raw materials are fermented in multiple rounds, the production cycle is one year, and the base Baijiu is stored for no less than three years.

曲坯（Qu billet）：Wuliangye uses wheat, barley and peas as raw materials, crushed, mixed with water, pressed into bricks, and cultivated under a certain temperature and humidity.

曲料粉碎度（Koji crushing degree）：Crushing degree of Daqu raw material.

酒母 (Jiumu): An expanded culture of yeast used in the brewing industry.

低温曲 (Low temperature koji): In the process of making koji, the highest temperature is less than 50℃ to make Daqu.

曲母 (Qumu): When making Wuliangye Daqu, inoculate a small amount of koji seeds.

堆积发酵 (Pile fermentation): In the process of making Jiang-flavor Baijiu, the grains are spread and dried, mixed with koji, and piled up into piles for opening fermentation.

翻仓 (Turn over): During the process of making Wuliangye, adjust the position of the blanks to make them evenly cultivated.

干曲仓 (Dry starter storehouse): Special room for storing koji.

磨曲机 (Grinding machine): In Wuliangye brewing, a machine that uses dry koji to grind.

磨曲 (Grinding): The operation of grinding koji with a koji grinding machine as required.

陈香曲 (Chen-aroma qu): High temperature koji made in August.

Lesson 3　Tasting

Objectives

At the end of this lesson, students should know:

1. The fastest method for sensory evaluation.
2. Four steps to Baijiu tasting.
3. The tasting of Nong-flavor style.

Text

Sensory evaluation is an important approach for Baijiu quality assessment. Baijiu judges, as the core technical power in distilleries, determine product quality and enterprise fate. Baijiu sensory evaluation means that the color, aroma and taste of Baijiu are scientifically analyzed and evaluated through the eyes, nose, mouth and other sensory characteristics and quality of Baijiu. This evaluation method has the characteristics of rapid, accurate, sensitive and simple (Table 2-1).

Evaluation is the main basis for determining the quality of Baijiu and blending and flavoring level. Without good evaluation skills, qualified Baijiu products cannot be drawn. The quality inspection of semi-finished Baijiu, the identification of storage level of incoming Baijiu, the quality control of blending and flavoring, and the qualified inspection of finished Baijiu are all inseparable from sensory inspection. At present, sensory evaluation is the most widely used and the most effective method for Baijiu inspection, and it is of great significance for Baijiu quality inspection in China.

Baijiu flavor training. There are 12 types of Chinese Baijiu, which the basic types are Jiang-flavor, Nong-flavor, Qing-flavor and Mi-flavor. The other types are derived from the four basic types. There are differences and connections between each type. How to distinguish the twelve fragrances? Baijiu tasters generally adopt the method of "first look, second smell, third taste and fourth square" to conduct different flavor Baijiu in different regions. Due to different production techniques, there are great differences in the types and quantities of flavor substances, thus forming different sensory characteristics. A good Baijiu taster not only needs to have excellent Baijiu tasting skills, but also needs to grasp the sensory characteristics of each flavor due to the different production processes.

Table 2-1   Common terms in Baijiu tasting

Item \ Fragrance	Nong-flavor style	Mild-flavour style	Jiang-flavour style
Color	Style colorless and transparent, slightly yellow, no suspended matter; turbid.	The same as Nong-flavor.	Slightly yellow, transparent, and the rest are the same as the Nong-flavour.
Aroma	Cellar incense is relatively strong and rich, and the compound aroma with ethyl acetate as the main body is pure and coordinated; cellar incense is insufficient and less pure, with the odor of pit mud.	The fragrance is pure and elegant, the compound fragrance with ethyl acetate as the main body is elegant and harmonious; the fragrance is not pure enough and has an individual fragrance.	The Jiang-flavor is prominent, elegant and delicate, and the empty cup lasts for a long time; the sauce is not prominent but has a strange fragrance, and the empty cup has a short fragrance.
Taste	Moderately sweet, mellow, harmonious, refreshing, and long aftertaste; spicy, peculiar and miscellaneous.	Moderately sweet, smoothly, refreshing and sweet; spicy, not refreshing enough, and miscellaneous.	Mellow, soft, fullness, long aftertaste; peculiar smell, miscellaneous smell, short aftertaste, ordinary.
Style	The style is outstanding, with a strong fragrance style, and typical; the style is average and the typicality is poor.	The style is outstanding, with a fragrant style and typical; the style is average and the typicality is poor.	The style is outstanding, with Jiang-flavor style and typical; the style is average and the typicality is poor.

Questions on Text

Compare the color of Nong-flavour style, Mild-flavour style and Jiang-flavour style.

Supplementary reading

Expanding the ranks of Baijiu tasters and improving the skill level of Baijiu tasters can not only guarantee the quality of finished Baijiu, but also guarantee the quality of original Baijiu and carry out reasonable grading storage, which is conducive to the development of different levels of Baijiu products and bring benefits to the enterprise. With the improvement of people's living standard, as a Baijiu production enterprise, the quality management should be strengthened constantly. To strengthen the analysis and research of the nutrients in Baijiu, carry out the analysis of the ingredients that are harmful to human health, and strive to develop new products that are beneficial to the public health. This requires enterprises increasing investment, actively cultivate and bring up large quantities of Baijiu tasting and Baijiu quality inspection and technical personnel, actively introduce advanced testing equipment, need more Baijiu tasting and inspection technology personnel in line with the responsible for the enterprise, highly responsible spirit, to people's health conscientious, give full play to the important role of professional technology.

The future

Currently, the pretreatment and analysis of liquor flavor component identification method is relatively mature, as the study of different kinds of liquor and the scientific research technology advances, more components will be digging, now more concerned about is whether these ingredients in liquor flavor play a key role, or some of liquor flavor characteristics is caused by what, what is the source of these components? The determination of some typical aroma characteristics and the source of some different aroma is still the dilemma that the major wine companies are facing. The study of the formation mechanism, production pathway, transformation pathway in the brewing process and the factors influencing the content of characteristic aroma components will provide scientific theoretical basis for the regulation of Baijiu aroma.

In the sensory analysis of Baijiu flavor components, the determination of the contribution of flavor components and the study of the interaction between flavor components will be helpful for scientific researchers to analyze liquor more comprehensively and systematically and solve the related problems of Baijiu flavor. The sensory evaluation of Chinese Baijiu and the construction of flavor wheel provide basic data and scientific guidance for the study of Chinese Baijiu flavor, and also provide consumers with visual and specific liquor sensory information. The relevant flavor-oriented research and consumer preference research will further guide the production and improve the flavor of the products, so as to adapt to the younger characteristics of the future Baijiu consumption market.

Sensory evaluation

1. "Soft"

The first is "soft", which mainly refers to the thick taste of Baijiu. Good brands generally do not appear to be very scratchy, but quickly integrate into our digestive system, become warm water, and spread quickly.

2. "Sweet"

The second is "sweet", which is mainly for good drinkers. We will have a feeling when we eat. If we don't give you any vegetables for a meal, just let you eat, not for a long time, you can taste the rice is actually very sweet. The same is true for good Baijiu. If you just drink big mouthfuls, I believe you will definitely feel that Baijiu is also very sweet.

3. "Smoothly"

The third is "smoothly", which is mainly non-sticky. Good Baijiu brands are generally very smooth and non-sticky, so many Baijiu companies tend to work hard on the net.

4. "Fragrant"

The fourth is "fragrant", especially for aroma-flavor Baijiu brands. The flavor has an important influence on the quality of Baijiu. Flavor type is also generally called a very important parameter for the purity index of Baijiu enterprises.

5. "Aroma"

The fifth is "aroma", which is also an important indicator of many Baijiu as a core selling point. The material and emotional values embodied by Baijiu are very beautiful. Therefore, many Baijiu shopping guides will use this strategy.

6. "Refreshment"

The sixth is "refreshment", which is similar to the purity, mainly for guiding the psychological feelings during the drinking process of Baijiu.

7. "Mellow"

The seventh is "mellow", the thick taste is pleasant. The thickness of liquor taste is also very attractive to long-term drinking consumers. Terminal shopping guides must learn the language communication skills for the target group. Thick taste means that the Baijiu tail is thicker and you can't lose your marbles.

8. "Pure"

The eighth is "pure". Since the distinction between grain Baijiu and blended Baijiu, the grain purity of Baijiu has also become a very important indicator for judging the quality of Baijiu. The pure of the Baijiu represents the authenticity and purity of the source of raw materials for Baijiu.

9. "Mellowness"

The ninth is "mellowness", which is mainly the taste of Baijiu. Mellowness means that the uniformity of Baijiu is relatively good, and the characteristics such as layering of low-grade Baijiu will not appear.

10. "Traditional"

The tenth is "positive", which mainly describes that the taste of Baijiu is more in line with the psychological needs of the target population, thus showing a more authentic and formal taste.

Lesson 4 Packaging

Objectives

At the end of this lesson, students should know:
1. Different materials for making Baijiu packaging.
2. The influence of the Chinese culture on Baijiu packaging.

Text

Extraction of cultural elements:

One of the design methods of the Chinese Baijiu packaging is to extract ethnical elements from traditional culture, such as finding inspiration from traditional Chinese paintings, extracting abstract ink paintings and woodblock prints from the overall shape, and combining western modern modeling concepts and designs. The new graphic visual image created by this technique is not limited to traditional design thinking, but can also find inspiration from traditional Chinese folk art. Chinese paper-cutting, New Year paintings and other folk arts have simple shapes and beautiful structures, containing profound national cultural heritage (Figure 2-8). The colors are bright, with a strong Chinese flavor; you can also explore Chinese calligraphy and seal cutting skills in depth. The word art is another business card of the Chinese culture, and the shaping of the word shape is very important to enhance the sense of space of the packaging. For international Baijiu packaging, the readability and space collocation of English characters should also be considered when designing Chinese fonts and it is in line with international typesetting principles and reading habits.

Figure 2-8 Folk art   Figure 2-9 Ceramic bottle&Glass bottle

Inheritance of craftsmanship:

The craftsmanship is reflected in the use of materials in traditional Chinese Baijiu packaging. Traditional Chinese Baijiu packaging is mostly based on natural materials, and materials such as ceramics, hemp, and wood are more common (Figure 2-9). Taking materials from nature can empathize with nature, reduce people's sense of isolation of packaging, and increase the richness of Baijiu packaging to provide consumers with more choices. Nowadays, all countries in the world are advocating the consumption and production mode of circular economy. People yearn for the packaging design that is natural and recyclable. The packaging of Baijiu entering the international market should be close to nature while protecting nature, reducing the consumption of nature, and achieving "product-product". The circular packaging mode avoids excessive packaging, which can better adapt to the design trend of the international market.

Environmentally friendly packaging:

Environmental packaging has become an important part of the national environmental protection movement. In order to win the "hearts" of consumers, many companies are developing environmentally friendly packaging, such as degradable packaging and recycled packaging. The use of environmentally friendly packaging can not only become a major selling point for enterprises, but also reduce consumers' sense of guilt when buying products. Leading environmentally friendly packaging through advanced technology has become a development trend.

The first is use of 3D optical transfer paper is environmentally friendly, degradable, and environmentally friendly; the second is low-cost, no positioning hot stamping process and hot stamping foil are required, which reduces the overall cost of packaging and printing; the third is anti-counterfeiting and beautiful, the same paper can be used. It integrates various optical patterns, supports accurate overprinting, has patent protection, and is more anti-counterfeiting.

Adopting back to basics design to guide the development of Baijiu packaging towards environmental protection. For example, the packaging of Yang He micro-molecule Baijiu uses environmentally-friendly "Kraft paper" primary colors with "green" printing, highlighting the original ecological and healthy product characteristics, appearing simple and calm, and embellished with a few colors, the overall design tone adopts micro-elements and micro-design (Figure 2-10). The concept of presenting information with the most concise layout and elements.

Figure 2-10　Packaging

Questions on Text

What are the characteristics of Baijiu packaging?

Words to Watch

traditional [trəˈdɪʃənl] adj. 传统的；习俗的；惯例的
ceramics [səˈræmɪks] n. 陶瓷制品；陶瓷器；制陶艺术；陶瓷装潢艺术；ceramic 的复数
hemp [hemp] n. 麻布；麻类植物；大麻
techniques [tɛkˈniːks] n. 技巧；技艺；工艺；技术；技能；technique 的复数
readability [ˌriːdəˈbɪlɪti] n. 可读性；易读性；可辨性
environmentally [ɪnˌvaɪrənˈment(ə)li] adv. 与环境有关地；在环境方面地
craftsmanship [ˈkræftsmənʃɪp] n. 手艺；技艺；精工细作
Chinese paper-cutting 中国剪纸
calligraphy [kəˈlɪɡrəfi] n. 书法；书法艺术

Phrases and patterns

be limited to 限于
be based on 基于
in order to 为了

adapt to 适应

Supplementary reading

After the creative design, the Baijiu still needs to be sold, so it is important to position consumers. According to different levels, each Sichuan Baijiu brand has different designs. The product analysis from the corresponding Baijiu series of different consumption levels can well understand the differences of consumption levels in choosing Baijiu culture. For instance, there are three major series corresponding to four groups of people in LUZHOU LAOJIAO (a kind of Daqu). The top series is National Pits · 1573 which is mainly aimed at high-end consumers. They pay more attention to the realm in Baijiu culture and the presentation of noble identity. Therefore, it is a customized product, with customized No. 1 national and Supreme as typical representatives. This kind of Baijiu is naturally particular about packaging, and its packaging design takes nobility as the core of creativity. Consumers can identify the advantages of the National Pits · 1573 from the packaging. Based on the analysis of the package of the national, the main color is red, which is an auspicious, festive and atmospheric color for Chinese people; the body of the bottle is made of red porcelain, which can also reflect the positioning of the national. Porcelain itself is a representative of China, and red is a favorite color of the Chinese people. The national itself is designed for the 60th anniversary of the national day. These elements and colors reflect that LUZHOU LAOJIAO. Respect for the country and the people, and its good wishes and ardent hopes of the future life. The packaging design of LUZHOU LAOJIAO is undoubtedly the inheritance and development of Chinese culture. In contrast, the Junyao series of the National Pits is carried forward and inherited from the decoration. It is made up of multi edge and multi side cutting, as if the diamond cutting surface that is recognized by modern people. In addition, the design of bottle finally displays the words of "Junyao" incisively and vividly. Since ancient times, China has a firm and unshakable spirit. When the Junyao series is used in business occasions, the transparent glass bottle is like the sincerity of cooperation, and the implication of the decoration is like the determination of cooperation. Such high-end customized Baijiu expresses the determination and sincerity in business, which is also a good cultural heritage. LUZHOU LAOJIAO is the most commonly selected series for middle-end consumers, who pursue more humanistic realm, pay more attention to quality and pursue more elegance. In the face of such consumers, the creativity in packaging design is to create culture with humanity and quality. For example, as the four gentlemen of ancient China, plum, orchid, bamboo and chrysanthemum have their own characteristics and are praised by Chinese poets. And they are cultural heritage. LUZHOU LAOJIAO series is also designed from the perspective of elegance. It shows the spirit of self-improvement through plum, orchid, bamboo and chrysanthemum. It also gives the image of being indifferent to fame and wealth, and elegance to the bottle body. As if people are drinking it, they were in an elegant atmosphere. It inherits and carries forward the qualities of being indifferent to fame and fortune, as well as elegance.

The future

Glass bottles will become the absolute mainstream, and high-quality ceramic bottles will occupy the market with prices above 1000 yuan. Because glass bottles have several advantages, one is that they can be quickly mass-produced and standardized, and have high efficiency; the other is that glass bottles have great advantages in expressiveness (Figure 2-11). The colorless and transparent Baijiu is made of crystal clear glass. The best deduction carrier can also reflect the delicate surface process effect in the shape. The glass material can also be recycled and reused. On the whole, the overall cost of the glass bottle is cheaper than the porcelain bottle. This is the core element. Ceramic bottles, Chinese have a ceramic complex since ancient times, and in ancient times, only the royal family could enjoy good porcelain ware, which also created the status of porcelain materials in the hearts of the Chinese people. Therefore, high-end Baijiu is matched with high-quality ceramic bottles. Complement each other.

Figure 2-11 Glass bottle

Products with a sales price of less than RMB 100 will gradually be sold in naked bottles, mainly through labels and Baijiu bottles to reflect the grade. The status quo of Baijiu packaging in developed countries can be used for reference; in the current recognition of many people, light bottle Baijiu is a very low-end Baijiu, and most of the products are priced from a few yuan to about 20 yuan, and light bottle is to some extent low-end Baijiu label. The reason for this status quo is that the price factor is preconceived, creating low consumer impressions; second, the product design lacks a very personalized design, and the product materials are also very rough. Combining these points, it will be formed for a long time. The impression of light bottle products.

There will be more and more personalized product packaging, and there will be a diversified phenomenon in packaging styles and forms to meet the growing personalized needs of consumers (Figure 2-12). In an era of full market competition, more products with different types and concepts will inevitably be produced, as well as relatively single and different forms and styles.

Environmentally friendly materials, processes, printing inks and other low-carbon, degradable and recycled Baijiu packaging will become the mainstream of the market. The country's strict enforcement of food safety will also force the safety improvement of packaging materials and processes. With the successive introduction of corresponding national policies, Baijiu packaging will gradually move towards an environmentally friendly and low-carbon sustainable development path.

Figure 2-12　Packaging

Lesson 5　Distillation and Maturation

Objectives

At the end of this lesson, students should be able to:
1. Understand the distillation method of liquor.
2. Understand the benefits of liquor storage.
3. Understand the changes of various substances in liquor storage.

Text

Distillation is a unit operation that uses the difference in volatility of components to separate liquid mixtures. Heating the liquid mixture or solid fermented spirit causes the liquid to boil, producing vapors that contain more volatile components than the original mixture, while the remaining mixture contains more volatile components, resulting in a partial or complete separation of the components of the original mixture. The components are partially or completely separated. There are many methods for liquid distillation of condensed steam, including simple distillation and rectification. In the production of liquor, ethanol and its associated flavor components are separated and concentrated from solid fermented liquor or liquid fermented mash to obtain liquor. The unit operation that contains many trace aroma components and alcohol content is called distillation. It belongs to the simple distillation method of distilled spirits: solid fermentation method, liquid fermentation method and solid-liquid combined string distillation method.

98%~99% of the ingredients in liquor are ethanol and water, which constitute the backbone of the liquor, and the other 1% to 2% of the ingredients are mainly composed of trace organic acids, esters, fusel alcohols, aldehydes, ketones, sulfur compounds, and nitrogen compounds and extremely small amounts of inorganic compounds (forms) etc., which determine the fragrance and taste of liquor, constitute the typicality and style of liquor. And the types of trace ingredients (or flavor substances and flavor ingredients) and the content of each trace ingredient in wine are closely related to the raw materials, production processes and storage and aging processes used in the production of liquor. The taste of newly distilled wine has a spicy, astringent, fragrant, bran and new wine smell, etc., but after a period of aging, the dryness and irritability of the wine is reduced, the wine is soft, the aroma is harmonious, the taste is mellow, and the aftertaste is long.

Liquor maturation is a continuation of the production process. It is a complex physical and chemical reaction process. During the maturation process, it is affected by changes in the external environment. Through changes in volatilization, intermolecular association, redox, esterification and hydrolysis, a variety of substances finally reach dynamic equilibrium.

During the maturation process, the esters undergo a hydrolysis reaction, which makes the esters have a significant downward trend, and the total acid shows an upward trend. Although the esters are hydrolyzed to produce alcohols, the low-chain alcohols are volatile. The downward trend is not obvious and tends to be stable.

Questions on Text

Why distillation is important in Chinese liquor brewing?

Words to Watch

distillation [ˌdɪstɪˈleɪʃən] n. 精馏；精华，蒸馏物
volatility [ˌvɒləˈtɪlɪti] n. 挥发性；易变；活泼

esterification [eˌsterifiˈkeiʃən] n. 酯化

Phrases and Patterns

solid fermentation method 固体发酵法
liquid fermentation method 液体发酵法
solid-liquid combined string distillation method 固液结合管柱蒸馏法
sulfur compound 含硫化合物
nitrogen compound 含氮化合物
inorganic compound 无机化合物
intermolecular association 分子间缔合

Supplementary reading

In the traditional solid-state fermentation method, the fermented and matured liquor is distilled by using a retort barrel. The liquor retort barrel is a cone with a top diameter of about 2m, a bottom diameter of about 1.8m, and a high lm conical distiller separated by a porous grate. The upper movable cover is connected with the cooler. The barrel is a unique distillation equipment that is different from other wine stills in the world. It is designed and invented according to the characteristics of solid fermentation wine. Since the advent of liquor, the distilling barrel has been used for thousands of years. After the founding of the People's Republic of China, with the substantial increase in production volume and technological transformation, the barrel has changed from small to large, the material has been changed from wood to reinforced cement or stainless steel, and the cooler has been changed from a sky pot to a straight tube type, which improves the cooling efficiency. The basic operation points of the gap-type manual charging remain unchanged, and the mechanization of continuous feeding and discharging has not been successful so far.

Alcohol distillation is based on the continuous addition of alcohol-containing fermented mash to the tower, so under constant steam heating conditions, the alcohol concentration of each layer of the tray is also constant, and various impurities at a certain alcohol concentration and temperature, according to different volatilization. The coefficient also has a specific content in each tray.

Lesson 6 Blending Technique

Objectives

At the end of this lesson, students should be able to:
1. Understand the Chinese Liquor blending process.
2. Understand the reasons for Chinese Liquor blending.
3. Learn how to blend Chinese Liquor.

Text

There are two ways to adjust wine. Chinese Liquor refers to the use of basic liquor and seasoning liquor fermented by solid grains, and does not use edible alcohol and food additives. "Alcohol" refers to the use of edible alcohol completely or in large proportion when liquor is blended, supplemented with a small proportion or completely without the use of basic wine or seasoned wine fermented by solid grains. How to distinguish between alcoholic liquor and alcoholic liquor. In the production process of three types of liquor, solid method, liquid method and solid-liquid method, the liquor of solid method liquor is hooked with liquor, while liquid method and solid-liquid liquor are alcohol.

For all solid-state liquor, after making Jiuqu, fermentation, steaming, aging, etc., it is essential to adjust. The solid-state production method is completely coordinated by the basic solid fermentation of grain and the seasoning wine. No edible alcohol and non-white wine fermentation-flavored substances can be added. This determines that the solid-state liquor must be mixed with wine.

The liquid method liquor is basically composed of edible alcohol and a small amount of grain-fermented solid method liquor. It is even completely based on edible alcohol. It does not contain grain-fermented solid method liquor at all. It is flavored and flavored with food additives. It is based on edible alcohol and a small amount of solid-state fermentation of basic wine, seasoning wine, etc., or entirely alcohol.

The solid-liquid liquor is composed of not less than 30% solid liquor and liquid liquor, so the solid-liquid liquor is tuned with edible alcohol and a small amount of grain solid fermentation base wine, seasoning wine and so on.

Questions on Text

Simple analysis of three methods of blending liquor.

Words to Watch

liquor [ˈlɪkə] n. 酒，含酒精饮料；溶液
edible [ˈɛdɪbl] adj. 可食用的
additive [ˈædətɪv] n. 添加剂，附加剂
proportion [prəˈpɔːʃən] n. 比例；部分；均衡
distinguish [dɪsˈtɪŋgwɪʃ] vt. 区分；辨别
essential [ɪˈsɛnʃəl] adj. 基本的；必要的；本质的；精华的
coordinate [kəʊˈɔːdnɪt] v. 调节，配合；使动作协调
mix [mɪks] v. （使）混合；配制；参与
seasoning [ˈsiːznɪŋ] n. 调味品；佐料

Phrases and Patterns

refer to 谈到，涉及
in large proportion 占很大比例
between... and... 在……和……之间
be essential to 对……至关重要
be composed of 由……组成
based on 以……为基础

Supplementary reading

When it comes to blending, you have to talk about evaluation, because evaluation is the basis of blending technology. Liquor tasting is also called tasting or appraisal. It is a tasting and appraisal technique to identify the quality of wine through the eyes, nose, mouth, and tongue according to the quality standards of various types of liquor. It is very important and plays a decisive role. It is fast, accurate, simple, convenient and applicable. It can determine the product grade in time, which is convenient for graded storage and quality classification. At the same time, you can also grasp the changes and aging laws of wine during storage, and understand the changes of different wine quality in storage. Tasting is also the only way to check the blending combination and the quick and sensitive effect of seasoning. It is also an important means to guide production, new product development and quality control. Only with a high level of wine tasting technology can a good product be prepared. Only on the basis of evaluation can the blending work be carried out.

Blending is the process of comprehensively balancing the same type of wine with different characteristics according to a unified and specific standard. Its purpose is to unify the wine quality and standard, so that the quality of each batch of factory wine is basically the same. The purpose of the wine quality is to make up for the shortcomings, develop strengths and avoid shortcomings, and form a wine body with unique characteristics. Blending is also conducive to the development of new products and enhance the vitality of enterprises. The principle of blending is mainly based on the balance of the absolute content of various trace flavor components in the wine and the ratio between them, so that the molecules of various trace flavor components are rearranged and associated, so as to achieve the wine design standard.

Unit 2 Brandy

Lesson 1 Introduction

Objectives

At the end of this lesson, students should be able to know:
1. What brandy is.
2. What the types of brandy are.
3. Main raw material of brandy.

Text

Brandy has more sources than any other class of distilled spirits made in the United States and may be divided into two classes, grape brandy and fruit brandy. American grape brandy is produced from grape wine and is distilled mostly in continuous stills. It is almost always artificially colored with caramel and aged generally in new plain white oak barrels. Fruit brandy is made more often from apples, but includes the distillate from many other fruits and berries. Fruit brandy is usually made in other states than California, which produces practically all the grape brandy of the United States. Fruit brandy is generally distilled in pot stills at a lower proof than grape brandy and is aged principally in charred barrels from which it obtains its color.

As far as can be determined there is a small amount of methanol in all authentic brandy; grape brandy contains an average of less than 0.05 percent and fruit brandy, an average of about 0.1 percent. Fruit brandy ages and develops congenerics and extracts color and solids at about the same rate as whisky and rum in the same sort of package. Grape brandy in the plain packages changes more slowly and more gradually.

French Cognac brandy is in a class of its own. It is clearly distinguished by a uniform and unique character not found in any other brandy. American grape, apple, and other fruit brandies are distinct from any other brandy in that they possess the unmistakable natural flavor and pleasant aroma of the fresh fruit from which they are distilled. This flavor is not lost but rather enhanced during natural aging.

Of the various groups of congenerics characteristic of brandy, the esters are of most importance and are generally more abundant in brandy than in any other distilled spirit.

Questions on Text

How to make brandy?

Words to Watch

source [sɔ:s] n. 来源
proof [pru:f] n. 证明；证据
authentic [ɔ:'θɛntɪk] adj. 真正的，真实的；可信的
congeneric [ˌkɒndʒɪ'nerik] adj. 同类的，同属的
package ['pækɪdʒ] n. 包装
unmistakable [ˌʌnmɪs'teɪkəbl] adj. 明显的；不会弄错的

Phrases and Patterns

be made in 制造于
be divided into 被分为
produce from 从……生产
an average of 平均是……
as far as 只要
be abundant in 富有……

Supplementary reading

The production of brandy in the United States is rapidly regaining its pre-prohibition importance, both as regards quantity and quality. In 1900 something over 1,000,000 proof gallons were made, and by 1914 production had increased to 3,750,000 gallons of taxpaid beverage brandy. During the prohibition period, legal production of brandy was reduced to an insignificant volume. Under special permits, however, two plants were allowed to make 25,000 gallons each per year for warehousing and aging, medicinal, and other non-beverage uses. After the repeal of prohibition 2,500,000 gallons of beverage brandy were distilled in 1933. By 1934 approximately seventy brandy distilleries were operating, mostly in conjunction with wineries.

A large amount of the post repeal brandy is now old enough to meet the U.S.P. age requirements for medicinal brandy and to be bottled in bond according to the United States internal revenue provisions. At present most domestic brandy is being bottled under the "taxpaid" regulations.

California, with over 80 percent of all the brandy made in the United States, is the largest producer in the world. Most of the California production is grape brandy, but important quantities are also distilled from apples, peaches, apricots, and other fruits in this state.

France is by far the leading exporter of brandy to the United States; its Cognac has an overwhelming lead over all other foreign types. Greece, the second largest exporter to this country, sends us only about one tenth as much as France. Still smaller amounts are imported from Spain, Germany, Italy, and several other countries.

Lesson 2   Distilled Technique

Objectives

At the end of this lesson, students should be able to:
1. Know what equipment is needed to distill brandy.
2. Have a brief description of the brandy distillation process.
3. Know the time required to complete a brandy.

Text

Most wine distillations (Cognac, Armagnac, and brandy) can best be described as extractive and reactive batch distillation processes. The ethanol concentration determines the volatility of distillate aroma compounds. At the same time, many wine compounds are subjected to chemical reactions at low pH and high temperatures within the still.

The key to the quality of brandy lies in the grape raw material, distillation, storage, blending technology, and distillation technology is particularly important. Only through high-level distillation technology can the aroma components in the original grape wine be extracted to the greatest extent and be collected in the original brandy. There are usually three types of brandy distillation. Traditional Charente pot distillation is often used for high-end brandy distillation. Tower continuous distillation is mainly used for ordinary brandy distillation. Skin residue pot distillation is mainly used for skin residue brandy distillation.

Distillation time

The first distillation should be carried out before the completion of fermentation to the end of March of the following year, in order to prevent the rancidity of the original grape wine caused by the warming of the weather. The second distillation should be completed before May

1st of the following year.

Distillation method

Charente pot distillation is a second distillation, and the original grape wine is distilled once to obtain the crude distillation brandy with a wine degree of 26 to 29 degrees; then the crude distillation brandy is subjected to secondary distillation, and the distillate is pinched to remove the wine. The middle distillate of 65 to 70 degrees is the original brandy. The use of double distillation can ensure the delicate aroma of the brandy and a high alcohol content.

The distillate ethanol concentration can vary widely during the process; therefore, how the still is operated will strongly influence both aromatic quality and productivity. Distillers simultaneously have to cope with complex process dynamics, the highly nonideal thermodynamic behavior of the mixture, multiple distillate fractions, coupled control variables, variations in wine composition, and also take into account changes in consumer preference.

Questions on Text

Why should the distillation of brandy be completed by May?

Words to Watch

ethanol [ˈɛθənɒl] n. 乙醇，酒精
temperature [ˈtɛmprɪtʃə] n. 温度；体温；气温
storage [ˈstɔːrɪdʒ] n. 存储；仓库；贮藏所
original [əˈrɪdʒənl] n. 原件；原作；原物；原型
delicate [ˈdɛlɪkɪt] adj. 微妙的；精美的，雅致的；柔和的
fraction [ˈfrækʃən] n. 分数；部分
consumer [kənˈsjuːmə] n. 消费者；用户，顾客

Phrases and Patterns

describe as 描述为
at the same time 同时
lie in 在于
carry out 进行，实行
in order to 为了

Supplementary reading

Stills now in use in California for different kinds of brandy vary from a few simple pots and worm stills with capacities as low as 10 gallons per hour to large sixty-chambered duplex continuous stills producing up to 350 gallons per hour. Most of the stills are continuous, and have capacities of 120 to 150 gallons per hour, of the types known as De Valle, Sanders, Krenz, Ergot, Hebert, Barbet, and others. At present one hundred and one continuous stills (mostly Krenz) are operating in California as compared with only about ten pot stills. Some special patented pomace stills are in use for producing grappa and fortifying brandy.

General opinion among experienced California distillers is that their commercial brandies should be distilled at about 155 to 175 proof in order to obtain the maximum bouquet, grape flavor, and other desired characteristics. Brandy distilled at higher proof is increasingly neutral and approaches ordinary alcohol (neutral spirits) in its general qualities. This type of brandy rarely improves much on aging.

Apple and other fruit brandies are generally distilled in smaller pot stills at less than 160 proof.

Receiving tanks for newly distilled beverage brandy, although sometimes of wood, are more frequently of copper or are tin-lined in order to avoid discoloring the brandy when the proof is reduced to 100~105 proof before the brandy is placed in wooden barrels.

Recent work on distilled spirits has focused either on aging and distillation system contributions to the final chemical composition of the spirit or on the relationships between this composition and the aroma and odor of the spirit. How distillation recipes affect spirit composition and the problem of adapting operation recipes to help distillers accommodate changes in consumer preference or cope with varying wine composition have received little attention though.

Lesson 3　Barrel and Aging

Objectives

At the end of this lesson, students should be able to:
1. Know why brandy is aged.
2. Know how to choose oak barrels.
3. Know how long the brandy usually need to be aged.

Text

In the brandy industry, oak barrels have been widely used to store brandy and to improve its quality. It is widely recognized that fresh distilled spirit is undrinkable for its harsh taste, pungent smell, and some possible harmful side effects. Therefore, fresh distilled spirit should be stored in oak barrel for many years in order to obtain attractive color, complex aroma as well as harmonious and comfortable mouth feel (Figure 2-13).

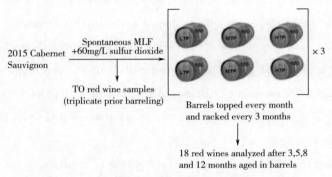

Figure 2-13 Oak barrel flavor

Experimental set up of red wine used and aged in oak barrels. LTP, low tannin potential; MTP, medium tannin potential; HTP, high tannin potential. 160, initial toasting temperature of 160℃; 180, initial toasting temperature of 180℃.

Extraction of oak wood compounds into brandy during barrel aging is of great importance since these can significantly modify the brandy's aroma as well as mouth feel. It is well known that geographical origin and species of oak tree and toasting of the barrel as well as seasoning and age of the barrel used all affect the quantity of compounds that can be potentially extracted from oak barrels by the brandy. The aging process of brandy in oak wood barrels involves several processes, that is, extraction of complex phenolic substances as tannins from wood, depolymerizing of structural molecules and their extraction to the spirits, and occurrence of physical and chemical reactions between oak wood components and spirits. These processes are very important for the final product quality such as taste, flavor, and color appreciation by consumers.

Aging in oak wood allows the extraction of a series of benzoic and cinnamic compounds to brandy. These compounds include vanillin, syringaldehyde, syringic acid, coniferaldehyde, gallic acid, and ellagic acid. Aging also brings about the modification of physical and chemical parameters of brandy. During aging, the transfer of different phenols and volatile compounds from oak wood into brandy depends on potentially extractable compounds that are initially present in the wood. Phenols are compounds in wine or brandy which influence the color, astringency, bitterness, oxidation level, and various physicochemical changes during its long aging period, while tannins are polyphenols that endow fresh wine astringency and bitterness, thus making it unpleasant to taste. However, appropriate amounts of these substances are desirable for full body and balanced structure of high-quality wine.

The USA and some European countries such as France are the main regions supplying oak barrels all over the world. In Europe, the most commonly used oak tree species are *Quercus robur* and *Quercus petrea*, and French oak barrel is rich in oak lactones, especially its cis-isomer compounds. On the other hand, the toasting degree of a barrel can also significantly influence the aromatic composition of brandy, which increases the content in volatile phenols, furfural compounds, vanilla as well as oak lactones.

Questions on Text

Please briefly describe the composition of brandy in oak barrels.

Words to Watch

undrinkable [ˌʌnˈdrɪŋkəbl] adj. 不能饮用的
pungent [ˈpʌndʒənt] adj. 辛辣的；刺激性的；苦痛的
molecule [ˈmɒlɪkjuːl] n. 分子，微粒
appreciation [əˌpriːʃɪˈeɪʃ(ə)n] n. 欣赏，鉴别；增值；感谢
cinnamic [sɪˈnæmɪk] adj. 含苯乙烯基的
physicochemical [fizikəuˈkemikəl] adj. 物理化学的
substance [ˈsʌbstəns] n. 物质；实质

Phrases and Patterns

be used to 习惯于
It is widely recognized that 人们普遍认为
as well as 以及
It is well known that 众所周知
such as 例如
rich in 富含
on the other hand 另一方面

Supplementary reading

Wine brandy is a complex mixture of hundreds of volatile compounds. However, the aroma research has led to the conclusion that only 5% of the volatiles identified in foods contribute to aroma and these compounds are called key odorants or odour-active compounds. Some studies have screened the key odorants in the aged wine brandies. Some odorant compounds result from the distillate, such as acids, alcohols, esters and terpenes while others originate from the wooden barrel, such as lactones, acids, aldehydes and phenols.

After the distillation, some distillates can remain from several months to many years in wooden barrels. This process, called ageing or maturation considerably modifies the odorant composition of the beverage and its sensory profile. Actually, several factors, such as the kind of wood, the heat treatment of the barrels and the ageing time have a significant influence on sensory and physicochemical composition of the brandies.

Such modifications are an outcome of several physical and chemical processes that take place in wooden barrels. Nishimura and Matsuyama classify these processes into seven types for all the matured distilled spirits: direct extraction of wood components; decomposition of macromolecules shaping the wood framework and its elution into the spirit; reactions between the wood compounds and the distillate compounds; reactions involving only the wood compounds; reactions involving only distillate compounds, evaporation of low boiling compounds and formation of stable clusters between ethanol and water.

Lesson 4　Blending Technique

Objectives

At the end of this lesson, students should be able to:
1. Know why should brandy be blended.
2. Know what base wine to use for brandy.
3. Know what the process of brandy blending is.

Text

Blended brandy is based on brandy of grape juice and blended with a certain amount of edible alcohol. Many countries in the world have blended brandy production. Among them, French brandy enjoys the highest reputation in the world. In addition to cognac, it also produces other grades and types of brandy. In France, the brandy is blended with edible grade molasses alcohol, blended with a certain proportion of grape distilled alcohol, brandy flavor and spices, and then aged in oak barrels for two years. France, the United Kingdom, etc. have all produced brandy flavors and fragrances, which are not only used domestically but also exported abroad.

The market sales volume of blended brandy depends on the quality of the product. If it is prepared with good quality, low cost, and appropriate price, there will still be a large market capacity. The key points of blending brandy production focus on the optimization and purification of edible alcohol, the selection and combination of brandy flavors, and blending technology. In recent years, according to the requirements of GB/T 11856-2008, the use of existing molasses wine.

The refined, corn alcohol and cassava alcohol have been studied on the optimization of the blended brandy process and the combination of spices to develop brandy products suitable for mass consumption.

Questions on Text

According to national standards, how to blend brandy?

Words to Watch

blend [blɛnd] v. 混合；交融；掺和
cognac [ˈkɒnjæk] 干邑白兰地
fragrance [ˈfreɪgrəns] n. 香味，芬芳
optimization [ˌɒptɪmaɪˈzeɪʃ(ə)n] n. 最佳化，最优化
purification [ˌpjʊərɪfɪˈkeɪʃən] n. 净化；提纯
requirement [rɪˈkwaɪəmənt] n. 要求；必要条件；必需品
combination [ˌkɒmbɪˈneɪʃən] n. 结合；组合；化合
consumption [kənˈsʌmpʃ(ə)n] n. 消费；消耗

Phrases and Patterns

base on 基于
in addition to 除……之外
depend on 依赖于
focus on 专注于

Supplementary reading

The classic after-dinner drink is also a popular anchor for cocktails.

Brandy evokes a certain prototypical image: the postprandial snifter, cradled by a well-heeled gentleman enjoying a nip in an elegant dining room. But that's the old school picture. Now that an increasing number of mixologists are building cocktails around the aromatic brown liquor, a new, broader audience is reaching for the spirit and as more than an after-dinner drink.

'Brandy is a hot mixer right now,' Dave Herlong, master mixologist for the Las Vegas based N9NE Group, says of the spirit, which is distilled from wine or other fermented fruit juices.

Beyond offering traditional and updated takes on classics such as sidecar and brandy Alexander, bartenders are stirring up renditions of regional favorites, including New Orleans' brandy milk punch (a brunch cocktail mixed with milk or half-and-half, vanilla extract, confectioners sugar and nutmeg) and Philadelphia Fish House Punch (which features regular and peach brandies, dark rum, sugar or simple syrup, lemon juice and, often, tea).

Cocktail lists also boast new brandy-based drinks. Custom House Tavern in Chicago serves the Almond Toast, which combines brandy, a homemade Marcona almond syrup and a few dashes of bitters. The Shanghai Martini at Karma, an Asian-themed restaurant in Mundelein, Ill., stars brandy, pineapple juice, cherry liqueur and blue curacao.

Unit 3 Whisky

Lesson 1 Introduction

Objectives

At the end of this lesson, students should be able to:
1. Briefly introduce what whisky is.
2. Know which countries have largely produced whisky.
3. Understand the origin of the word "whisky".

Text

Whisky (or whiskey, depending on where it hails from) is one of the world's leading spirits. Its history is every bit as distinguished as that of cognac and, like the classic brandies of France, its spread around the world from its first home-the Scottish Highlands, in whisky's case-has been a true testament to the genius of its conception. Tennessee sour mash may bear about as much relation in taste to single malt Scotch as Spanish brandy does to cognac, but the fact that they are all great products demonstrates the versatility of each basic formula.

Whiskies are produced all over the world now. As the name is not a geographically specific one, they may all legitimately call themselves Whisk (e) y. In Australia and India, the Czech Republic and Germany, they make grain spirits from barley or rye that proudly bear the name. The five major whisky-producing countries are Scotland, the United States, Ireland, Canada and Japan, which are covered in this chapter. The name "whisky" itself is yet another variant on the phrase "water of life" that we have become familiar with in the world of spirits. In translation, the Latin aqua vitae became uisge beatha in the Scots branch of Gaelic and usquebaugh in the Irish; it eventually was mangled into the half-anglicized "whisky" and was in official use by the mid-18th century.

In countries that lacked the warm climate for producing fermented drinks from grapes, beer was always the staple brew and, just as brandy was the obvious first distillate in southern Europe, so malted grains provided the starting-point for domestic production further north. Unlike brandy, however, which starts life as wine, whisky doesn't have to be made from something that would be recognizable as beer. The grains are malted by allowing then, to germinate in water and then lightly cooking them to encourage the formation of sugars. It is these sugars on which the yeasts then feed to produce the first ferment. A double distillation by the pot still method results in a congener-rich spirit that can then be matured—often for decades for the finer whiskies—in oak barrels.

Just as with other spirits that haven't had the life rectified out of them, whisky is nearly always truly expressive of its regional origins and the raw materials that went into it. For that reason, a passionate connoisseurship of this spirit has arisen over the generations, similar to that which surrounds wine. Even more than brandy, whisky handsomely rewards those who set out with a conscientious approach to the tasting and appreciation of the spirit.

Questions on Text

Briefly introduce how whisky is made.

Words to Watch

versatility [ˌvɜːsəˈtɪlɪti] n. 有多种用途，多功能性
geographically [ˌdʒɪəˈgræfɪkəli] adv. 地理学上
rye [raɪ] n. 黑麦
mangle [ˈmæŋgl] vt. 撕烂；损坏
germinate [ˈdʒɜːmɪneɪt] v. 发芽
congener [ˈkɒndʒɪnə] n. 同种的物；同类的人；adj. 同种的，同类的
connoisseurship [ˌkɒnəˈsɜːʃɪp] n. 鉴赏能力
conscientious [ˌkɒnʃɪˈɛnʃəs] adj. 勤勉认真的；一丝不苟的

Phrases and Patterns

every bit as 同样
all over the world 全世界
Czech Republic 捷克共和国

Supplementary reading

Both Scotch whisky and American whiskies were influenced by tax policy. An aggressive tax strategy at the end of the eighteenth century suppressed the manufacture of traditional whisky in the highlands, while encouraging lowland distillers to produce spirits that were far less desirable. In America the tax pressure seems to have promoted the practice of barrel aging, encouraging distillers to manufacture potable goods that did not require rectification.

Food adulteration concerns were global in the nineteenth century, and the legal response in the United Kingdom and in America were necessarily similar since global trade was affected. In both nations, highly distilled alcohol was being mixed with more traditional whisky, and marketed without distinction from the traditional product. Counterfeit spirits comprised of artificially flavored pure alcohol were also being manufactured. Public hearings on these subjects tended to conflate these two problems, and legal actions were hampered by there being no appropriate definition of 'whisky.' The rhetoric in the American debates was comically absurd on both sides. In the end, definitions of whisky were crafted that reserved particular classes for traditional products (e.g., single malt scotch and straight rye whisky) while permitting more highly rectified products to use the name whisky (e.g., single grain scotch whisky and light whisky). The naming conventions are subtle and not meaningful to most consumers.

Neither traditional highland scotch nor traditional American whiskies were originally aged by the distiller, and a typical public house consumer was likely to get unaged whisky diluted or modified by the publican. Even at the beginning of the twentieth century, some distillers sold their product by the barrel and not by the bottle, and many consumers had little knowledge of the provenance of their whisky.

This facilitated a number of abusive practices including the marketing of what today would be considered counterfeit goods.

The blending of highly rectified whisky with more traditional whisky was not, however, an entirely misleading practice. Consumer tastes in both Scotland and America evolved to favor more lightly flavored whisky, and the use of column spirits offered a cost advantage in principle. Mixing relatively pure alcohol with traditional whisky also reduced bluing, or chill haze formation, which consumers associated with impure products.

A significant difference between the Scottish case and the American case concerns authenticity. Single malt scotch whisky is significantly similar to its traditional antecedent. The most significant changes are barrel maturation in recent times, and the occasional use of oats and spices in the past. In contrast, the flavor profile of American whisky arose from practices during the Civil War which were not used at all one hundred years previously. After the Civil War, whisky was distilled with live steam and aged in charred new oak barrels. Pot distilled American whiskies of the eighteenth century were highly prone to charring, and were therefore rectified, flavored with prunes or peaches, artificially colored, and sold unaged.

THE FUTURE

Whisky is a drink attractive to discerning adults that has developed into epicurean status.

The appreciation of whisky has spread all over the world. It has been found in locations as diverse as Antarctica and the Sahara and is currently being distilled in the International Space Station (ISS) (a no-gravity environment). Many countries, such as India, have developed their own distinctive brands (Chapter 4), and this trend will continue, with each country establishing their own standards, regulations, and specific criteria to protect exclusivity. Ross Aylott has provided an excellent list of these criteria in Chapter 14 of this book. Taxation will always be with us, and new challenges will continue to impact the industry; for example, expanding populations will put pressure on the industry when the increased demand for food, water, and energy conflicts with development of potential new markets for whisky (and other beverages).

Lesson 2 Distilled Technique

Objectives

At the end of this lesson, students should be able to know:

1. What factors can affect the physical process of separation during distillation.

2. What chemical reactions can occur when the wash or low wines are heated for distillation.

3. The brief definition of "Distillation" in Whisky.

Text

Distillation is a process that achieves chemical separation by taking advantage of phase change. In the case of spirits, we are removing ethanol from water by leveraging ethanol's preference, relative to water, for the vapour phase. This can be shown in a temperature vs. mole fraction phase diagram.

Other reactions that occur during distillation include the decarboxylation of ferulic acid to 4-vinylguaiacol, and the formation of β-damascenone from neoxanthin. Maillard reactions, including the formation of furfural, are important in wash stills that are directly fired. That practice is largely obsolete, however, with closed or live steam being preferred.

In batch distillation using pot stills, the selection of the cut points in the still run has a marked influence on the flavour of the product. Generally, aldehyde and short-chain ester concentrations are determined by the primary cut from foreshots to spirit, while the concentration of fusel alcohols and acids is determined by the cut from spirit to feints. Other factors, which affect the physical process of separation during distillation, are still design parameters such as the still height, the angle, and length of the lyne arm and the type of condenser used. In addition, a number of chemical reactions can occur when the wash or low wines are heated for distillation. This is particularly true for batch distillation in pot stills where the heat impact is greater. Examples include the formation of sulphur compounds from sulphur-containing amino acids, the break-down of unsaturated fatty acids to carbonyl compounds, and the dehydration of β-hydroxypropionaldehyde to acrolein. Copper catalyses many of these reactions.

In continuous stills, congener concentrations are determined by the design, operation, and efficiency of the still. The range of reactions described for pot stills can still occur but become progressively less important as distillate strength increases. Collection of spirit at high ethanol concentrations (>90%) results in low levels of congeners in the spirit.

Questions on Text

What factors can impact the congener concentration in continuous stills?

Words to Watch

leveraging [ˈliːvərɪdʒɪŋ] n. 杠杆作用；杠杆

Maillard reaction 美拉德反应

foreshot [ˈfɔːʃɒt] n. （蒸馏酒时）初馏分，酒头

vapour [ˈveɪpə] n. 蒸气；潮气；雾气

condenser [kənˈdɛnsə] n. 冷凝器；电容器

acrolein [əˈkrəʊlɪɪn] n. 丙烯醛

Phrases and Patterns

pot still 壶式蒸馏器

lyne arm 连接臂，莱恩臂

Supplementary reading

Distillation of grain whiskey spirit at the Midleton Distillery uses a three-column system, combining a beer column with a two-column extractive distillation unit. This system is illustrated in Figure 2.2. Fermented beer at 13.5% abv enters the beer column just above halfway (tray 22 of 37) onto the top tray of the stripping section. Steam introduced at the base of the column (via a reboiler) strips both alcohol and congeners as it travels up the column, bringing the alcohol-enriched vapour to the top of the column, where it is condensed and drawn off to produce beer-column high wines with a typical strength of 72% abv.

The extractive distillation column operates on the basis that dilution water added at the top of the column changes the volatility of the higher alcohols, aldehydes, and esters, which will now travel to the top of the column as the diluted alcohol stream flows down the column. As the diluted alcohol stream flows down, it reaches an area of the column known as the pinch, where the optimum congener and alcohol concentrations occur. At this point, the alcohol stream, now at approximately 20% to 22% abv, is transferred directly to the rectifying column.

At the top of the extractive column, the congener-rich vapour is condensed to produce a headstream that flows through a decanter to allow for separation of the fusel oils. The fusel oil stream is decanted and the remaining liquid is transferred back to the top of the extractive column as reflux. The diluted alcohol stream, also containing congeners not removed in the extractive column, enters the rectifier at tray 22 and is concentrated back up the column using steam entering at the base of the column. As the ethanol and congener stream travels up the column, a side stream is taken off between trays 56 to 62, where a majority of the remaining higher alcohols, particularly isoamyl alcohol, concentrate. This side stream is returned to the top section of the extractive column.

Near the top of the column, at tray 74, the product stream is removed, giving a spirit with a typical strength of 94.4% abv. The overheads from the top of the column are condensed to provide reflux, with the remainder being recycled back to the top of the extractive column. Bottoms from both the rectifier column and the extractive column are used to maintain a level in the dilution water tank with the excess going to drain.

The expansion of the Midleton Distillery requires grain whiskey production to increase from the current 23 million LAs to 42 million LAs per year. This expansion is currently underway, and the three-column unit described above will be replaced with a new system capable of delivering the required capacity. Although the process streams previously described will essentially remain the same, there will be one significant change: the introduction of a multi-pressure distillation system that will deliver very significant energy savings. Trials in the extractive distillation pilot plant have shown that a beer column operating under vacuum coupled to a rectifier operating with an over pressure will give normal distillate quality while delivering energy savings in the region of 50%. The extractive distillation column will operate close to atmospheric pressure and will utilise mechanical vapour recompression (MVR) technology. This new system was commissioned toward the end of 2013.

Lesson 3 Barrel and Aging

Objectives

At the end of this lesson, students should be able to:
1. List changes in barrelled whisky chemistry over time.
2. List some maturity sensory descriptor.
3. Draw a diagram of congener changes during maturation of whisky.

Text

Changes in barrelled whisky chemistry over time include:
- the reaction of ethanol with wood—selective lignin ethanolysis;
- the extraction of substances from barrel wood;
- oxidation (moderated by transport through barrel wood);
- reactions in solution: esterification, modification of tannins;
- volatilization, especially of sulphur compounds; and sorption.

In new charred oak, extraction occurs rapidly, i.e., in the first 2 years. Ester formation is much slower, progressing in a linear fashion over at least 8 years. Oxidation is also slow, as measured by the slow decrease in pH after the fast extractive period. Early wet chemistry established this general picture about 100 years ago, with modern techniques filling in details of chemical speciation and reaction mechanism.

The most striking conclusion of this work is that very little change appears to happen after a few years. However, at 3 years spirit is barley eligible to be called whisky in many jurisdictions, and few consumers would consider it to be "mature". Maturity in the sense of age can be measured in terms of slow esterification and oxidation reactions. Maturity in the sense of smoothness, roundness, and other sensory descriptors is not easily understood through the picture outlined here. One approach to understanding these sensory characteristics of maturity is through links between maturation chemistry and human physiology, as exemplified by recent work by Aoshima, Koga, and colleagues. Other approaches, concerning the structure of the spirit, are explored in the following chapter. As yet, however, patience is the only proven approach to making a whisky smooth.

During the maturation period, the new distillate becomes highly modified as a result of its contact with the cask (Figure 2). The composition of the cask wood has a major effect on the profile, which can be recreated batch after batch with minimal variation. Blending is most commonly associated with blending single-malt whiskies and single-grain whiskies together to form blended Scotch whisky. However, the principles of blending are also crucial to the spirit quality of other types of Scotch whisky, such as single malts.

The role of the blender:

The blender is generally considered to be the sensory authority of a whisky company. He or she is the person who designs recipes for new blends or malt expressions, manages the inventory to ensure the sustainability of the existing and future brand portfolio, and who passes sensory judgement on all stages of the Scotch whisky manufacturing process. The exact scope, nature, and indeed the title of the role can vary depending on the company, such that the size and complexity of the company may dictate additional responsibilities for the blender, such as new product development; scientific, technical, and regulatory matters; ambassadorial duties; and internal and external education. The main element is, of course, management of the inventory and development and maintenance of the wood policy to ensure that the spirit profiles of the product portfolio remain the same and that future blends are consistent with the "house style" of the brand. The blender is, in essence, the "Guardian of the Whisky" and should pay particular attention to anything that could impact its flavour—in the distillery, the cask, the warehouse, the blend centre, or the bottling hall; during transit to the market; and, finally, off trade and on trade in the hands of the consumer. Thus, the knowledge and experience of the blender must be extensive.

In order to comprehend blending and indeed the role of the blender more fully, it is important to have a deep understanding of Scotch whisky and how it is defined in law. It is of critical importance that the blender is aware of, and understands, the law concerning Scotch whisky and operates strictly within its parameters.

Questions on Text

What is blending most commonly associated with?

Words to Watch

batch [bætʃ] n. 一批；一炉；一次所制之量

recipe ['resəpi] n. 配方，处方

inventory ['ɪnvəntri] n. 存货，存货清单；详细目录

scope [skəʊp] n. 视野；范围；余地

dictate ['dɪkteɪt] v. 指示；命令

Lesson 3 Barrel and Aging

Objectives

At the end of this lesson, students should be able to:
1. List changes in barrelled whisky chemistry over time.
2. List some maturity sensory descriptor.
3. Draw a diagram of congener changes during maturation of whisky.

Text

Changes in barrelled whisky chemistry over time include:
- the reaction of ethanol with wood—selective lignin ethanolysis;
- the extraction of substances from barrel wood;
- oxidation (moderated by transport through barrel wood);
- reactions in solution: esterification, modification of tannins;
- volatilization, especially of sulphur compounds; and sorption.

In new charred oak, extraction occurs rapidly, i.e., in the first 2 years. Ester formation is much slower, progressing in a linear fashion over at least 8 years. Oxidation is also slow, as measured by the slow decrease in pH after the fast extractive period. Early wet chemistry established this general picture about 100 years ago, with modern techniques filling in details of chemical speciation and reaction mechanism.

The most striking conclusion of this work is that very little change appears to happen after a few years. However, at 3 years spirit is barley eligible to be called whisky in many jurisdictions, and few consumers would consider it to be "mature". Maturity in the sense of age can be measured in terms of slow esterification and oxidation reactions. Maturity in the sense of smoothness, roundness, and other sensory descriptors is not easily understood through the picture outlined here. One approach to understanding these sensory characteristics of maturity is through links between maturation chemistry and human physiology, as exemplified by recent work by Aoshima, Koga, and colleagues. Other approaches, concerning the structure of the spirit, are explored in the following chapter. As yet, however, patience is the only proven approach to making a whisky smooth.

During the maturation period, the new distillate becomes highly modified as a result of its contact with the cask (Figure 2). The composition of the cask wood has a major effect on the extent of this modification and varies with the species of oak used, the pretreatment applied to the wood, and the previous use of the cask. In general, American oak (*Quercus alba*) yields higher concentrations of vanillin and cis-oak lactone, while Spanish oak (*Quercus robur*) gives higher concentrations of tannins. Charring increases the levels of lignin breakdown products and lactones extracted during maturation, whereas previous use of a cask greatly reduces the amount of extractable lignin breakdown products and lactones. The use of sherry casks results in raised levels of tannins and sugars. The cask is the major determinant of mature quality, but other factors such as filling proof, climate, and warehouse conditions also affect the course of maturation. Higher temperatures produce greater evaporation rates and also increase extract levels and some reaction rates, but do not necessarily produce a more mature product.

Questions on Text

What is the function of charring?

Words to Watch

ethanolysis [eθəˈnɒləsɪs] n. 乙醇分解
sorption [ˈsɔːpʃən] n. 吸附作用；吸收作用
smoothness [ˈsmuːðnəs] n. 平滑；柔滑；平坦
pretreatment [ˈpriːˈtriːtmənt] n. 预处理
vanillin [vəˈnɪlɪn] n. 香草醛
lactone [ˈlæktəʊn] n. 内酯
char [tʃɑː(r)] v. 烧焦

Phrases and Patterns

at least 至少
chemical speciation 化学形态

Supplementary reading

Whiskey maturation at the Midleton Distillery follows a strict wood management policy, for both pot still whiskey and grain whiskey. As with whisky distilleries in Scotland, the United States is the main supplier of oak casks, which come via two cooperages in Kentucky. Selected distillery run barrels from a number of bourbon distilleries are checked, repaired if necessary, and then shipped to Midleton on a regular basis throughout the year. Current requirements are in the region of 140,000 first fill barrels (B1s) per year.

Midleton's maturation policy requires 40% of pot still spirit to be matured in B1s, while grain whiskey typically utilises 60% B1s in its maturation profile. This approach for grain whiskey maturation would not be typical for a Scotch grain whisky, where barrels will normally have performed a series of malt fills before being used for grain whisky maturation. All barrels from the United States are shipped as standing barrels, with no shooks being imported.

The use of sherry-seasoned casks also plays a significant role in the maturation of whiskey at Midleton Distillery. In this case, sherry butts (500 L) are commissioned from a specific cooperage in Jerez de la Frontera, Spain, which uses only European oak (*Quercus robur*). They are then seasoned with Oloroso sherry for two years in three bodegas within the Jerez appellation. The company commits to the wood with the cooperage two years before manufacture, to allow for an 18-month air-drying period. All of the wood is sourced from the Galicia region of Northern Spain. In addition to sherry-seasoned oak butts, smaller quantities of fortified wine casks are sourced from Portugal (Port), Sicily (Marsala), Malaga (Malaga wine), and Madeira (Madeira wine).

All casks (barrels and butts) are held in palletised warehouses during the maturation period. There are six barrels per pallet, and a stack is seven pallets high; there are four butts per pallet, going four high. All barrels remain on the pallet at all times, and they are both filled and emptied through the head. This significantly reduces the handling and rolling of barrels, which only come off the pallet for repair or when they are being culled from the population. Typical use of a barrel would be three maturation cycles before being culled. Currently, no cask rejuvenation is practised at the Midleton Distillery or in any other Irish distillery.

Lesson 4 Blending Technique

Objectives

At the end of this lesson, students should be able to:
1. State the purpose of blending in Scotch Whisky.
2. Know what the responsibility of a blender is.
3. Tell the factors that can impact the flavour of whisky that blenders need to know.

Text

In Scotch, the purpose of blending is to create a Scotch whisky with a particular flavour profile, which can be recreated batch after batch with minimal variation. Blending is most commonly associated with blending single-malt whiskies and single-grain whiskies together to form blended Scotch whisky. However, the principles of blending are also crucial to the spirit quality of other types of Scotch whisky, such as single malts.

The role of the blender:

The blender is generally considered to be the sensory authority of a whisky company. He or she is the person who designs recipes for new blends or malt expressions, manages the inventory to ensure the sustainability of the existing and future brand portfolio, and who passes sensory judgement on all stages of the Scotch whisky manufacturing process. The exact scope, nature, and indeed the title of the role can vary depending on the company, such that the size and complexity of the company may dictate additional responsibilities for the blender, such as new product development; scientific, technical, and regulatory matters; ambassadorial duties; and internal and external education. The main element is, of course, management of the inventory and development and maintenance of the wood policy to ensure that the spirit profiles of the product portfolio remain the same and that future blends are consistent with the "house style" of the brand. The blender is, in essence, the "Guardian of the Whisky" and should pay particular attention to anything that could impact its flavour—in the distillery, the cask, the warehouse, the blend centre, or the bottling hall; during transit to the market; and, finally, off trade and on trade in the hands of the consumer. Thus, the knowledge and experience of the blender must be extensive.

In order to comprehend blending and indeed the role of the blender more fully, it is important to have a deep understanding of Scotch whisky and how it is defined in law. It is of critical importance that the blender is aware of, and understands, the law concerning Scotch whisky and operates strictly within its parameters.

Questions on Text

What is blending most commonly associated with?

Words to Watch

batch [bætʃ] n. 一批；一炉；一次所制之量
recipe ['resəpi] n. 配方，处方
inventory ['ɪnvəntri] n. 存货，存货清单；详细目录
scope [skəʊp] n. 视野；范围；余地
dictate ['dɪkteɪt] v. 指示；命令

ambassadorial [æmˌbæsəˈdɔːrɪəl] adj. 大使的；使节的

portfolio [pɔːtˈfəʊlɪəʊ] n. （公司或机构提供的）系列产品，系列服务

critical [ˈkrɪtɪkl] adj. 关键的，临界的，决定性的

Phrases and Patterns

flavour profile 香味剖析

minimal variation 最小变量

manufacturing process 生产过程

Supplementary reading

THE BLENDERS' CHALLENGE

Earlier in this chapter, it was stated that a blended Scotch whisky is a blend of one or more malt whiskies with one or more grain whiskies. Usually, the reality is that a great many more malts and a variety of grains are used to make a blended Scotch whisky. The malt and grain profiles are distinctly different from one another, due to differences in the method of production. The grains tend to have a flavour profile that contribute light and clean characteristics to the blend. However, it would be folly to presume that different grains can be interchanged in a blend, as each grain make has its own character and interacts in a different way with the malt component; therefore, a wrong selection could result in an unacceptable difference in the blend. The malts, on the other hand, exhibit a variety of sometimes very different flavour characteristics and have higher levels of congeners. The flavour profiles of the resulting malt and grain blends are dependent not only on the malts and grains chosen but also on the exact percentage of each of the makes. Therefore, given that the main objective of the blender is to create a consistent blend from batch to batch, no matter the inventory situation, it is important to create a system of categories or classes in order to make the inventory more manageable. The first level of category is generally "malt" and "grain", and the percentage that each type contributes to the blend is usually the top line of the recipe. The next level of category is generally based on the flavour profile of the malts and the grains. The exact nature of these categories varies from company to company, and this is not information that is readily shared in the industry. However, it generally involves grouping makes of Scotch whiskies together based on common organoleptic characteristics. The use of categories or classes, as they are sometimes known, means that, in practical terms, if a particular make is not available on the day of blending then a substitute make can be selected from its category in order to maintain the flavour profile of a particular blend. These categories are again assigned a percentage based on their contribution to the recipe. The next level contains the individual makes of malts and grains and their percentage contributions to the particular blend. Another important aspect of the blend is, of course, the casks in which the individual components of the blend have been matured. The wood profile, governed by the wood policy, for individual blends is a key element of the recipe that must be adhered to, so the recipe must not only conform to the individual makes and their percentages but also fall within the specifications for the individual wood categories. Bringing all of these elements together on the day of blending requires careful planning!

NEW PRODUCT DEVELOPMENT AND INNOVATION

New product development (NPD) is the process through which ideas for new products are assessed and subsequently developed into physical products for the marketplace. The aim of the NPD process is always to launch a successful product onto the market in the shortest possible time. The NPD process varies from company to company, but generally the process starts with a perceived need from an individual market, or a category of the global market, depending on the size and product range of the company. For the liquid part of the idea, a liquid brief is generated that is shared with the blender and other relevant parties. The information contained in the brief, as a minimum, should include the alcoholic strength and the age, as well as any claims or unique selling points that the market wants to claim on the label. The blender will then engage with the market representatives to ensure that all aspects of the brief are clearly understood by all parties. The blender will then check that all of the attributes fall within the Scotch Whisky Regulations of 2009 and that the inventory is sufficient to cover the anticipated demand. The next step is to proceed to create test samples that the blender considers matches the liquid brief. The test liquids are then assessed both internally and by the market through a series of blind sensory assessments; that is, panellists assess the liquid without prior knowledge of the samples presented. At this stage, a preferred liquid may have been identified, in which case the sample is officially approved by the market, rendering the liquid brief "frozen", which means that no more changes to the brief are permitted without impacting on the launch date.

If the new liquid is also required to be prepared for bottling in a different manner or packaged in material that may be new to the brand, the blender will be involved in testing to ensure that the new conditions do not have a detrimental impact on the flavour profile of the product.

Unit 4 Other Distilled Alcoholic Beverages

Lesson 1 Vodka

Figure 2-14 Vodka

Objectives

At the end of this lesson, students should be able to know:
1. The apparent characteristic of Vodka.
2. Why Vodka has affordability.
3. Why Vodka becomes the perfect postmodern drink.

Text

Stripped to its essence, vodka (Figure 2-14) is little more than pure alcohol distilled from grain-a clear liquid without colour, odour or taste. Yet, thanks to its remarkable potency and versatility, it has become the world's favourite libation. Vodka can be taken neat in one gulp à la Russe, or sipped as a cocktail when flavoured by almost anything palatable. Vodka can even provide the base for a refreshing drink, when combined with fruit juice, soda water, tonic or ginger beer.

One distinct advantage vodka has over its rival spirits is, to paraphrase an advertisement, that it leaves one breathless-which is a coy way of saying that it leaves no smell of liquor on one's breath. For this reason (and the fact that it looks like water), vodka is often said to have a 'clean' appeal, and is reputedly the drink of choice among tippling bartenders the world over.

Vodka also has another inherent advantage over other alcoholic drinks: affordability. Its raw materials are relatively cheap and plentiful (cultivating fields is far less labour-intensive than tending vineyards), and its distillation process is comparatively quick, simple and efficient. Moreover, vodka knows no vintages; there is no elaborate ageing process—after all, why would anyone bother to age a drink that is basically without taste?

Another advantage is that vodka is extremely stable and non-perishable. A Russian joke underscores this point: a drunkard enters a liquor store and asks, "Do you have any fresh vodka today?"

The indignant merchant snaps, "What do you mean 'fresh', you idiot? It's vodka."

"I am no idiot," the customer retorts. "I drank two of your bottles yesterday and they made me sick!"

Yet for all vodka's inherent simplicity, this enticing elixir defies a simple description: vodka is many things to many people. Even when taken pure, some experience no taste at all, while others detect subtle flavours imbued by the basic ingredients. Even these connoisseurs probably do not experience exactly the same sensations on their palates.

Present-day vodka makers are keenly aware of vodka's widespread appeal and marvellous marketing possibilities. They design, package and pitch their brands to target a variety of markets including women, youths, macho types, bon vivants, connoisseurs and adventurers. Consumers, for their part, proudly proclaim their identities by their choice of brand. Vodka is, in fact, the perfect postmodern drink: marketers carefully construct a brand image while consumers define themselves—in no small part—by the brand they choose.

Most imbibers turn to vodka for pleasure, comfort, warmth, courage, consolation or even inspiration. Some maintain that it has healthful applications. As an alcoholic drink, however, it is easily abused and can be potentially a fearful, destructive force. Paradoxically, some drink vodka to heighten their appreciation of life's pleasures, while others use it to suppress life's pains, be they physical or emotional. Alas, it appears that one must use vodka either one way or the other, for no one can simultaneously enjoy life while seeking to escape from it.

Questions on Text

Why Vodka is the drink of choice among tippling bartenders the world over?

Words to Watch

strip [strɪp] n. 带；条状物

essence ['esns] n. 本质，实质；精华；香精
libation [laɪ'beɪʃən] n. 饮酒；奠酒祭神仪式
potency ['pəʊtənsi] n. 效能；力量；潜力；权势
rival ['raɪvəl] n. 竞争对手 vt. 比得上
inherent [ɪn'hɪərənt] adj. 固有的；内在的；与生俱来的，遗传的
affordability [əˌfɔːdə'bɪlɪti] n. 支付能力；负担能力
pitch [pɪtʃ] n. 球场；程度
consolation [ˌkɒnsə'leɪʃən] n. 安慰；慰问
ingredient [ɪn'griːdiənt] n. 原料；要素；组成部分
paradoxically [ˌpærə'dɒksɪkəli] adv. 自相矛盾地；似是而非地；反常地

Phrases and Patterns

without colour 没有颜色的
thanks to 幸亏
combine with 与……结合
fruit juice 果汁
soda water 苏打水
ginger beer 姜汁啤酒

Supplementary reading

We will start with a couple of recipes for vodka, as it is both the most popular spirit and the simplest to make. Originating in Russia, vodka is generally considered to be odourless, colourless, and flavourless. In other words, it is supposed to be as close to neutral in character as possible. Because vodka is supposed to be flavourless and therefore is not supposed to carry any flavours across during the distillation process, it does not matter what you ferment to make vodka. The control is in the distillation process. Vodka is generally distilled using a reflux method, as this allows for the purest spirit, although some vodkas are produced by distilling multiple times using pot distillation.

Because vodka is neutral, it is extremely versatile, which is likely the reason for its popularity. Being neutral in character also makes it ideal for flavouring, making it very useful in the production of other types of liquor and liqueurs. Both home distillers and some commercial distilleries will use vodka or NGS (neutral grain spirits) to produce products such as gin, whiskey, and rum, as well as liqueurs such as coffee liqueur, amaretto, and Irish cream. In fact, many home distillers are more than content staying with a simple sugar wash, which will be distilled to high purity, carbon filtered, and flavoured using essence, and they may never venture out from there. This should not be frowned upon by those who wish to be more involved in the process. This is a hobby, and each person is limited by a different level of free time and interest. To start, we will repeat the simple sugar wash, followed by an all-grain vodka recipe.

Lesson 2　Rum

Figure 2-15　Rum

Objectives

At the end of this lesson, students should be able to know:
1. what grade Rum has.
2. what the critical characteristic of Rum is.

Text

For many drinkers, rum (Figure 2-15) can be an afterthought: a little boost to turn a glass of juice into a nice cocktail and help you justify sitting in a thatched bar listening to Jimmy Buffett. The truth is, though, that rum can be as satisfying on its own as a glass of whiskey. You just have to choose wisely.

Like all liquors, rum is a spirit distilled from sugar. Like most modern liquors, rum has a handful of characteristics that make it rum. But within those parameters, there are variations ranging from light to dark to amber to spiced. Point is, rum's complicated.

Rums are produced in various grades. Light rums are commonly used in cocktails, whereas "golden" and "dark" rums were typically consumed straight or neat, iced ("on the rocks"), or used for cooking, but are now commonly consumed with mixers. Premium rums are made to be consumed either straight or iced.

The flavour of rum can be influenced by many factors—the characteristics of the strain of sugar cane, the age at which it is harvested, the purity of the molasses, how many times the molasses is distilled and to what proof, and how the resulting alcohol is aged. This allows vast latitude for an accomplished producer to create different effects, and for at least 400 years distillers and cellar masters have experimented and honed their skills.

Questions on Text

What factors can impact the flavour of Rum?

Words to Watch

afterthought ['ɑ:ftəθɔ:t] n. 随后添加的东西，追加物
boost [bu:st] v. 促进；宣扬 n. 改善；推动
cocktail ['kɒkteɪl] n. 鸡尾酒
spice n. 香料 v. 添加香料

Phrases and Patterns

a handful of 一把，一小撮
range from 从……到……

Supplementary reading

The islands of the Caribbean, Puerto Rico, Dominica, US Virgin Islands, British Virgin Islands, Martinique, Saint Lucia, Saint Martin, Sint Maarten, Saint Vincent, Antigua, Grenada, Saint Kitts, Barbados, Guadeloupe, Trinidad, and Marie Galante, sound like a roll call of West Indian holiday destinations. With Demerara on the continent of South America, they contribute over 230 bottled rum products to the consumer market.

From the early days of the sugar estates, dependent on slave labour, the products, both bottled and bulk, in cask or larger container, found their way to the four corners of the earth as bulk exports for maturation in the USA, Canada or Europe or as famous bottled products, produced and bottled in the Caribbean.

So varied are the individual rums that they can be used as mixers or like cognac or fine malt whiskies, as post prandial drinks to savour after a good meal.

Blenders of rum can reflect on its nautical past with some apprehension when pirates were ambassadors of the drink, while Churchill's famous dictum that the British Navy ran on 'Rum, sodomy and the lash', sends shivers down the spine (The Rum Information Bureau).

It was Admiral Penn, who first issued rum to sailors, sweetened with limes, in 1655. In 1731, the Navy Board introduced a daily ration of half a pint per rating. Between 1731 and 1740, so many sailors had plunged to their deaths from the rigging of ships, that in 1740, Admiral Vernon, noted for wearing a coat of grogram, commanded that the rum be issued 50 : 50 diluted with water. Thus a mixture of rum and water was henceforth known as 'grog'.

Admiral Nelson, following his death, at the Battle of Trafalgar was transported to England in a container of rum. On the return, a shortage of rum ensued and one or two tars helped themselves to a tot from the cask, which obviously contained droplets of Nelson's blood. Thus rum also earned the appellation, 'Nelson's Blood'.

In 1970, the Admiralty Board abolished the daily ration.

From the blood and sweat of the early slaves, stretching back three hundred and fifty years, to the abolition of slavery and the emancipation of the slaves in the former colonies, just over one hundred and sixty years ago, rum continues to be perfected by freemen, descendants of the original slaves and by East Indians to the present day.

Modern equipment and techniques are replacing some of the age-old traditions but the quality of the product remains superb.

Rum, like whisky, cognac and armagnac, can be deemed to be a noble spirit in the tradition of cask matured products.

As long as the world continues to demand cane sugar, rum will continue to be distilled wherever the cane is grown.

Lesson 3   Tequila

Figure 2-16   Tequila

neutral [ˈnjuːtrəl] adj. 中立的，中性的

juniper [ˈdʒuːnɪpə] n. 杜松；桧或刺柏属植物

consequent [kɒnsɪkwənt] adj. 随之发生的，作为结果的

drunkenness [ˈdrʌŋkənəs] n. 醉态；酒醉

Phrases and Patterns

in the presence of 在……面前；在……存在下

according to 根据

water down 加水冲淡

a type of 一种

compete with 竞争

Supplementary reading

Yesterday and Today

More than any other spirit, gin can incite an argument or elicit poetry. Whisky brings to mind the highlands of Scotland and whiskey, the dusky vales of Ireland; rum connotes pirates and the triangular trade; vodka speaks of steely Russian politicos and stark Siberian winters. All of these spirits have their own fascinating histories, but gin's history is truly a world history charting a path from the Middle East to Europe to America.

Since ancient times, gin's central component, juniper, has been used as a curative in the cultures of Egypt, Greece and Rome. During the time of the bubonic plague, juniper was the supreme panacea across Europe. When genever was developed in Holland, it became both a currency and a daily ration for the Dutch East India Company, whose members took it to places as diverse as Argentina and Indonesia.

In Great Britain, gin crossed social boundaries as the tipple of both the impoverished and the aristocrat. Like the Dutch, the British East India Company took gin with them as they colonized, spreading it to India and beyond. In America, gin-whether genever, Old Tom, London Dry or 'bathtub'-was intrinsic to the birth and evolution of the cocktail, more so than any other spirit. Modern gins from all corners of the globe-Sweden, New Zealand, America, Spain-speak to an evolving international spirits culture that embraces innovation and experimentation.

One look at the past decade suggests that gin has reclaimed its central position in the drink world. Certainly, it offers a variety of styles unlike any other spirit. Traditional London Dry appeals to the classicist, who embraces its juniper-laden character. Vodka-lovers and those looking for something that makes them rethink gin will be rewarded by a foray into the new world of modern botanicals. And for those eager to take a journey back in time, the mellow sweetness of Old Tom or the whiskey-like intensity of oude genever provide a glimpse into and taste of drink history.

Regardless of one's tipple, gin requires a palate that embraces flavour, not a body that merely desires inebriation. It is a spirit whose rocky evolution parallels mankind's history, both lowly and noble. In the world of ardent spirits, gin is the most alchemical of all, able to transform two humble elements—grain and juniper—into an elixir of infinite complexity and reward.

Chapter 3
Blended Alcoholic Beverages

Unit 1   Liqueur
Unit 2   Other Blended Alcoholic Beverages

Unit 1   Liqueur

Lesson 1   Fruit Liqueur

Objectives

At the end of this lesson, students should be able to:
1. Understand where the word liqueur came from.
2. Know what legal regulations are for liqueur.

Text

Fruit spirits and liqueurs, in particular, are some of the most popular spirit-based beverages made from fruits. The choice of raw material for their production depends on the climatic conditions in the producer's country, the storage method and the harvest time. Fruits such as, plums, cherries, melons, apples and pears are most frequently used in the production of fruit spirits. The type and form of fruit and the production method (fermentation, distillation and maturation process) have a huge significance in relation to the quality of the final product. As a result scientists continuously research the complex composition of fruit spirits. High-performance liquid chromatography is used to monitor the fermentation process, while GC coupled with MS or FID is employed for analysing the volatile fraction. Spectrometric analysis is applied to determine the contents of selected compounds in fruit spirits. Moreover, a sensory evaluation is still used to assess the organoleptic properties of these beverages. On-going research on fruit spirits enables monitoring the content of compounds that negatively influence human health. Such research also improves the quality control process, product authentication and the identification of botanical and geographical characteristics of fruit spirits in relation to fruits used in their production.

Liqueurs are another group of alcoholic beverages made from fruits, amongst other things. The name liqueur comes from the Latin word liquefacere, which means 'to melt' or 'to dissolve'. Liqueurs are made via concurrent dissolution or mixing of a number of components. According to legal regulations, liqueurs are colourless or coloured sweetened spirit-based beverages produced by flavouring ethanol or the distillate of agricultural origin. The minimum ethanol content in liqueurs is 15% volume, while the sugar content, expressed as inverted sugar, in most cases equals 100 g/L of liqueur. However, there are numerous traditional liqueurs that contain 35%~45% of ethanol. Products of agricultural origin, such as herbs (roots, seeds and flowers), fruits (whole fruit, peel and stones) and fruit juices, as well as other food products, inter alia, dairy products and wines, are used in the production of liqueurs. Moreover, essential oils, and natural and synthetic aromas are also used. There are three possible ways to obtain natural extracts, that is, infusion, percolation and distillation. The production of natural extracts begins with the soaking of the selected product (e. g. fruit) in ethanol at 40~60℃ for a couple of days. Next, the filtered solution undergoes maceration, followed by the distillation process. Most frequently, liqueurs are coloured by adding, inter alia, caramel and honey. The most popular liqueurs are made of apple, cherry and plum.

Questions on Text

State the ways to obtain natural extracts.

Words to Watch

cherry ['tʃɛri] n. 樱桃；樱桃树
moreover [mɔ:r'əuvə] adv. 而且；此外
spectrometric [ˌspektrəu'metrik] adj. 光谱测定的；分光仪的；能谱仪的
authentication [ɔ:ˌθɛntɪ'keɪʃ(ə)n] n. 证明；鉴定；证实
percolation [ˌpɜ:kə'leɪʃən] n. 过滤；浸透
chromatography [ˌkrəumə'tɒgrəfi] n. 色谱法，色层分析法，层析法
plum [plʌm] n. 李子；梅子
melon ['mɛlən] n. 甜瓜；瓜
infusion [ɪn'fju:ʒən] n. 浸出法，浸提
undergo [ˌʌndə'gəu] vt. 经历，经受

Phrases and Patterns

in relation to 关于
fermentation process 发酵过程

Supplementary reading

Constituents of Possible Concern

Constituents present in fruit spirits that may be seen as potentially having a negative influence on health include methanol, hydrogen cyanide and ethyl carbamate. In this context, however, it should be noted that their potential for harm is generally minimal, compared with that of alcohol itself.

Methanol is formed during the production process by the enzymatic hydrolysis of fruit pectin. The scheme of this reaction looks as follows:

An increase in methanol concentrations is influenced by, amongst other things, the content and level of methylation of pectins and the activity of the original pectin methylesterase in the fruit. Methanol is contained in all alcoholic beverages. Its concentrations are usually at a low level, however, in the case of fruit spirits produced together with stones and skins, the content of this compound may reach up to 1000 g/hL of 100% volume alcohol, and even more. According to EU directives for plum spirits, the concentration of methanol must not exceed 1200 g/hL of 100% volume alcohol.

Cyanogenic glycosides are a natural ingredient, but they are also contained in fruit stones. Stones may become damaged during mash preparation and cyanogenic glycosides from the stones may come into contact with enzymes in the fruit mash. As a result, cyanogenic glycosides are then degraded to hydrocyanic acid. Prolonged storage of the fermented mash may lead to the release of hydrocyanic acid from intact stones. Small amounts of hydrogen cyanide have an advantageous effect on the aroma of alcoholic beverages. In larger amounts, however, it can be harmful to the body. Regulation EC 110/2008 states that the HCN content should not exceed 7 g/hL of 100% volume alcohol.

Ethyl carbamate is a potential genotoxic carcinogen in humans. This compound is formed during the fermentation of food products. It has been proven that the highest concentrations of ethyl carbamate occur in spirit-based beverages made of stone fruits. In certain environmental conditions, such as exposure to light and high temperatures, ethyl carbamate can be formed from various substances, including hydrogen cyanide, urea and citrulline, which are present in food and beverages. The highest ethyl carbamate content is observed in the final stage of fermentation, when the ethanol concentration peaks. The allowed ethyl carbamate content in fruit spirits is 400 μg/L.

Research on methanol, HCN and ethyl carbamate is described in a further part of this publication.

Origins of Flavour

The quality of spirits is influenced by a natural aroma of fruits used for their production. The general aromatic profile of fruits is influenced by many factors: the geographical origin, the method of cultivation, storage and time of harvest. Fruits contain several hundred compounds, which have a cumulative influence on the general aroma. In many cases, fruit aroma is influenced by a mixture of several compounds, which do not show such properties individually, for example, hexyl 2-methylbutanoate, hexyl acetate and ethyl hexanoate contribute to apple aroma. In many cases, some compounds occur in many kinds of fruits and they are not a distinguishing feature of one aroma, for example linalool, nonanal, (E) -2-hexenol and limonene. Additionally, aromatic compounds are characterized by a different aroma depending on the concentration of a given compound, for example eugenol, which has a fruity aroma at lower concentrations, and a pungent aroma at higher ones. The botanical origin, that is, the difference in the composition between various cultivars of one kind of fruits, also makes it more difficult to determine the general aromatic profile of fruits, for example a compound from the terpene group, α-farnesene, was identified only in plums from the *P. salicina* and *P. domestica* cultivars but was not detected in the *P. cerasifera*, *P. ussuriensis* and *P. spinosa* cultivars.

Lesson 2 Eggnog Liqueur

Objectives

At the end of this lesson, students should be able to:
1. Get the key points of making German Eggnog liqueur.
2. Understand what Dutch eggnog liqueur is.

Text

Homemade German Egg Liqueur (Eierlikör)

Egg liqueur is widely used in Germany as either an ingredient in cakes or drizzled on top. It would make a wonderful gift to someone who enjoys baking. It's also a great gift for anyone who likes eggnog or rich and creamy beverages. When giving as a gift, be sure to label it to keep refrigerated. I also recommend using yolks from pasteurized eggs.

How does German Egg Liqueur taste? Think about super concentrated eggnog without nutmeg. Eierlikör is very thick with a consistency similar to sweetened condensed milk. It is typically served in cordial or old style flat-like champagne glasses so you could lick the last of it out of the glass if you were so inclined. Of course, glass licking is up to you. It may not be great etiquette but it sure is fun.

Ingredients:

1 vanilla bean, 10 egg yolks from pasteurized eggs recommended, 1 to 1/2 cups superfine sugar, 1 cup heavy whipping cream, 3/4 cup sweetened condensed milk and 1 cup light rum.

Instructions:

1. Cut vanilla bean lengthwise and scrape out the seeds. Discard outer bean. 2. Place egg yolks, vanilla seeds, and sugar in the bowl of a stand mixer fitted with the whisk beater. Beat on high speed for 10 minutes. 3. Slowly add cream and condensed milk and beat for 7 minutes. 4. Slowly add rum and beat for 3 minutes. 5. Pour egg liqueur in decorative bottles or containers. Seal and refrigerate for up to 3 weeks.

Recipe Notes:

Homemade German Egg Liqueur (Eierlikör) is a rich, sweet, and decadent beverage. It can also be used in baking or drizzling on cakes.

In Dutch speciality, there is a liqueur called Advocaat or advocatenborrel. It is essentially a customized version of the humble eggnog, without the milk: a mixture of simple grape brandy with egg yolks and sugar, as thick and as yellow as tinned custard. Most of it is sold in this natural form, although it is possible in the Netherlands to buy vanilla and fruit-flavoured versions. As a result of its velvety texture and bland wholesomeness, advocaat is often thought of as a drink for the elderly, and is commonly added to mugs of hot chocolate or strong coffee.

There are a few widely available brands of advocaat on the market: the red-labelled Warninks is probably the most familiar, but Fockink, and the liqueur specialists, Bols and De Kuyper, also make it. The standard bottled strength is quite low for a liqueur-around 17% ABV, which is about the same strength as the average fortified wine.

Questions on Text

How to define Advocaat?

Words to Watch

champagne [ʃæm'peɪn] n. 香槟酒

refrigerate [rɪ'frɪdʒəreɪt] v. 冷藏；使冷却

nutmeg ['nʌtmɛg] n. 肉豆蔻

cordial ['kɔːdiəl] adj. 兴奋的；热忱的，诚恳的　n. 甜香酒，甘露酒

etiquette ['etɪket] n. 礼节，礼仪，规矩

pasteurize ['pɑːstʃəraɪz] v. 用巴氏灭菌法给……消毒

lengthwise ['lɛŋθwaɪz] adv. 纵向地；纵长地

scrape [skreɪp] v. 刮掉；削去

vanilla [vəˈnɪlə] n. 香草　adj. 香草味的
yolk [jəʊk] n. 蛋黄
custard [ˈkʌstəd] n. 蛋奶沙司；蛋挞

Phrases and Patterns

be sure to 必定
similar to 类似
up to 高达
condensed milk 炼乳
beat on 拍打
natural form 自然形态
as a result of 作为……的结果
egg yolk 蛋黄

Supplementary reading

Eggnog can trace its roots back as far as the 14th century, when medieval Englishmen enjoyed a hot cocktail known as posset. Posset didn't contain eggs—the Oxford English Dictionary describes it as "a drink made of hot milk curdled with ale, wine, or the like, often sweetened and spiced" —but over the years eggs joined in on the festive fun.

While the egg-laden version of posset was popular with the English, it became less common as time went by. Milk and eggs were both scarce and expensive, and the sherry and Madeira used to spike the mixture was pricey, too. Over time, the concoction became a drink that only aristocrats could really afford.

All of that changed in the American colonies, though. What we lacked in parliamentary representation we made up for in easy access to dairy products and liquor.

Since many Americans had their own chickens and dairy cattle, tossing together a glass of eggnog was no problem, and the drink's popularity soared among the colonists even as it sagged back home.

This disparity in the drink's popularity on either side of the pond endures to this day; eggnog's popularity in the U. K. still lags far behind its holiday ubiquity here in the States. In fact, here's how the Guardian's Andrew Shanahan memorably described the drink in 2006: "People rarely get it right, but even if you do it still tastes horrible. The smell is like an omelette and the consistency defies belief. It lurches around the glass like partially – sentient sludge." Appetizing!

Using Your Noggin

The word "eggnog" itself has fairly murky origins, but many etymologists think the name stems from the word "noggin," which referred to small wooden mugs that were often used to serve this type of drink. Others propose a similar story but explain that the "nog" comes from the Norfolk slang nog to refer to the strong ales that were often served in these cups. Still others think the name is a contraction of colonial Americans' request to bartenders for an "egg-and-grog" when they wanted a glass.

Furthermore, while the drink itself may date back to medieval times, the word "eggnog" is a relatively recent invention. The first recorded instances of the use of "eggnog" only date back to the late 18th century, and by that time, bartenders in the young United States had already tweaked the recipe to give it a more American twist. The Madeira and sherry that English aristocrats had used for their version of eggnog were scarce on this side of the pond, but we had plenty of rum and whiskey. In 1800, author Isaac Weld, Jr. described the American recipe for eggnog as consisting of "new milk, eggs, rum, and sugar, beat up together."

Have One for George Washington

Yes, early Americans loved their eggnog, and you can use this fact to your advantage if you down a few too many glasses this year. Simply point out that you're in good company with the likes of George Washington. Kitchen records from Mount Vernon indicate that Washington served an eggnog-like drink to visitors, and since the general wasn't strapped for cash, he didn't skimp on the sauce. Washington's potent recipe included three different types of booze: rye whiskey, rum, and sherry. Nobody could tell a lie after having a few cups of that.

Not everyone had Washington's funds, though. A thorough look at historical recipes reveals that for most tipplers, the type of booze they snuck into their nog didn't really matter as long as there was something to give it a little kick. In addition to rum, ale, whiskey, and wines, an 1879 collection of recipes from Virginia housewives features a recipe that calls for 12 eggs, eight wine-glassfuls of brandy, and four wine-glassfuls of wine. Another calls for three dozen eggs, half a gallon of domestic brandy, and another half-pint of French brandy. Something's telling us these shindigs got a little wild.

You Might Not Want to Read the Label

If you pick up a carton of commercial eggnog at the supermarket, you're probably getting much more nog than egg. FDA regulations only require that 1.0 percent of a product's final weight be made up of egg yolk solids for it to bear the eggnog name. For "eggnog flavored milk," the bar is even lower; in addition to requiring less butterfat in the recipe, this label only requires 0.5 percent egg yolk solids in the carton.

Of course, there are other good reasons why we don't tip back eggnog year-round. Sure, nobody's reaching for a nice cup of something custardy on a hot day, but it's not very good for you at all. A relatively small four-ounce cup of store-bought eggnog boasts a whopping 170 calories (half of them from fat), nearly 10 grams of fat, and over 70 mg of cholesterol. (If you're keeping score at home, that's around a quarter of your recommended daily intake of cholesterol.)

If you choose to eschew these commercial brands in favor of mixing up your own eggnog, you'll probably want to use pasteurized eggs to suppress the risk of a nasty case of salmonella; not even a playlist featuring "Do They Know It's Christmas?" ruins a holiday party quite as quickly as making everyone in attendance violently ill. Don't use unpasteurized eggs under the old argument of "the booze will kill the germs," either. The FDA advises that this strategy isn't likely to work.

Lesson 3   Coffee Liqueur

Objectives

At the end of this lesson, students should be able to know:
1. What kind of liqueur can be the alcohol base to make coffee liqueur.
2. What method can be used to accelerate the aging of coffee liqueur.

Text

Coffee liqueur is a spirit-based liqueur flavored with coffee and the historical fable of its origins dates it to the 1700s. Even though coffee liqueurs all share a common ingredient, the taste of each one varies based on the spirit used and whether the product contains added sweeteners or flavors, such as vanilla. Rum is a common alcohol base used to make coffee liqueur, although tequila, brandy or vodka can also be a main ingredient. Rice wine is an eastern alcoholic beverage made from rice, originating from China. It is a very popular alcoholic drink for eastern people; not only is it used for drinking but is also greatly used in Chinese cuisine and in other Asian cuisines. Rice wine became a very important necessary for the daily life of eastern people. However, no coffee liqueur beginning with rice alcohol has been reported so far.

Coffee liqueur made from rice wine base was treated by ultrasonic wave for accelerating the aging process. After the coffee liqueur was treated by ultrasonic wave for 6 h, the color was able to reach the level of conventionally aged coffee liqueur at 120 days, as well as being able to maintain a high degree of lightness. The alcohol content, caffeine content and turbidity were found similar with 180 days of conventional aging; the content of main volatile compounds viz. butanol, isoamyl alcohol, ethyl hexanoate and ethyl lactate achieved the similar level as well as furan-methanol content which achieved 89.2% of furanmethanol level in conventional aging for 180 days. The aroma and taste are similar to the results of conventional aging for 180 days. Ultrasonic wave treatment is a good alternative for rice-wine-based coffee liqueur to shorten the duration of aging (Table 3-1).

Aging is one of the most important processes to improve the quality of wine, spirit and alcoholic beverage. Complex chemical reactions involving sugars, acids and phenolic compounds can alter the aroma, color, mouthfeel and taste in a way that may be more pleasing to the taster during aging. In conventional aging, coffee liqueur takes at least 6 months to reach the quality marketable level. It would be desirable to be able to accelerate this process, so that the same character of coffee liqueur could be produced in a much shorter time period. The ultrasonic wave treatment provided a method of aging that resulted in coffee liqueur that is aged as short as 6 h having the character conventionally achieved only after 6 months of aging.

Table 3-1 Changes of color, alcohol, caffeine and turbidity in rice-wine-based coffee liqueur treated by ultrasonic wave

	Time (h)			
	0	2	4	6
L	2832b	2839b	2850ab	2870a
a	26.15c	26.62bc	26.95ab	27.55a
b	18.30b	18.70b	19.05ab	19.85a
Alcohol/%	29.93a	29.43b	29.13c	29.13d
Caffeine/ (mg/L)	218.41b	232.40ab	244.74a	246.15a
Turbidity (A660)	$12.57e^{-2}$b	$13.24e^{-2}$ab	$13.90e^{-2}$ab	$14.43e^{-2}$a

Questions on Text

What is the benefit of ageing?

Words to Watch

hexanoate [ˈheksənəʊt] n. 己酸盐
furan [ˈfjʊəræn] n. 呋喃；氧杂茂
cuisine [kwiːˈziːn] n. 烹饪；风味；菜肴
ultrasonic [ˌʌltrəˈsɒnɪk] adj. 超声的 n. 超声波
butanol [ˈbjuːtənəʊl] n. 丁醇，同分异构酒精
isoamyl [aɪsəʊˈæmil] n. 异戊基
ethyl lactate n. 乳酸乙酯
methanol [ˈmɛθənɒl] n. 甲醇

Phrases and Patterns

necessary for 对……是必要的
ultrasonic wave 超声波
isoamyl alcohol 异戊醇

Supplementary reading

Caffeine Content

Caffeine content in coffee liqueur increased as the ultrasonic wave treatment time increased. After ultrasonic wave treatment for 6h, the caffeine content increased from 218.41 to 246.15 mg/L. Caffeine content of conventionally aged coffee liqueur increased with time during 180 days of aging. From day 0 till day 90 the caffeine content in the liqueur increased rapidly and after 90 days the increasing rate slowed down and became stable; there was no significant change ($P < 0.05$). The increase of caffeine content may be caused by residual of fine coffee particles in the liqueur. During the aging process, ethanol acts as an extraction solution and caffeine was gradually released. Ultrasonic wave treatment caused rapid increase of caffeine in the coffee liqueur. Coffee liqueur that underwent ultrasonic wave treatment for 4h has similar amount of caffeine as conventional aging for 90 days.

Ultrasonic extraction has been widely used in the extraction of various compounds due to the relatively low cost and simple instruments needed. The use of high-power ultrasound in the extraction of functional and bioactive components of plant materials has been reported by many researchers. It has been shown that ultrasound largely improves the extraction rate by disrupting plant cells and hence increasing the diffusion of the cell contents across the cell wall. The beneficial effects of sound waves on extraction are attributed to the formation and asymmetrical collapse of microcavities in the vicinity of cell walls leading to the generation of micro-jets rupturing the cells. Ultrasound power produces its effects via cavitation bubbles. These bubbles are generated during the rarefaction cycle of the wave when the liquid structure is literally torn apart to form micro bubbles which collapse in the compression cycle and the pressures of hundreds of atmospheres and temperatures of thousands of degrees are generated during the collapse of the bubbles. It is also thought that the pulsation of bubbles causes acoustic streaming which improves mass transfer rate by preventing the solvent layer surrounding the plant tissue from getting saturated and hence the enhancement of convection. It was inferred that due to ultrasound treatment, the temperature of coffee liqueur and the pressure in liqueur increased as well as the mass transfer improved, which caused the accelerated release of caffeine from those fine particles of coffee.

Ethanol and water mixtures are commonly used for the extraction of water soluble compounds from plant materials. In green tea, minimum caffeine value was obtained at the lowest ethanol concentration and reached the maximum value at 70% of ethanol concentration in the ultrasonic extraction time of 60 min. However in this study, the changes of alcohol content among ultrasonic-wave-treated coffee liqueurs were different significantly but negligible in amount. The change in the alcohol content of coffee liqueur should not take responsibility for the increase of caffeine both in ultrasonic wave treated and conventionally aged coffee liqueurs.

Ultrasonic wave treatment of coffee liqueur for 6h reached 98% of caffeine content of conventionally aged coffee liqueur at 180 days, therefore ultrasonic wave treatment accelerates aging as caffeine are released, which shortens aging time.

Lesson 4 Anisette

Objectives

At the end of this lesson, students should be able to:
1. Know what the author think the differences between anis and pastis.
2. Understand the medical function of Anis extracts.
3. Find out where the metals in Anisette come from.

Text

Anis

CONFUSION REIGNS as to the precise differences between anis and pastis, and indeed whether there are any meaningful differences at all. They are both flavoured with the berries of the aniseed plant, originally native to North Africa, and are popular all around the Mediterranean. They both turn cloudy when watered, and are both claimed as the respectable successor to the outlawed absinthe.

One august authority claims that pastis should be flavoured with liquorice rather than aniseed, although the two are very close in taste.

Another claims that anis is simply one of the types of pastis. Still another claims that, whereas anis is a product of the maceration of aniseed or liquorice in spirit, pastis should properly be seen as a distillation from either of the two ingredients themselves.

They can't all be right of course, but for what it's worth, I incline to accept the last definition. For one thing, anis tends to be lower in alcohol than pastis—liqueur strength rather than spirit strength. The one thing we can be sure of is that pastis is always French (the word is old southern French dialect), whereas anis—particularly with that spelling—can also be Spanish. In Spain, there are sweet and dry varieties, whereas French anise tends mainly to be dry.

Ever since the days of the medical school of Salerno, and probably earlier, extract of anis has been seen as a valuable weapon in the apothecary's armoury. It is thought to be especially good for ailments of the stomach.

Anisette

Anisette is quite definitely a liqueur. It is French, Sweetened, and usually somewhat stronger than anis. The most famous brand is Marie Brizard, from the firm named after the Bordelaise who, in the mid-18th century, was given the recipe by West Indian acquaintance.

Anisette is obtained by maceration and distillation of alcoholic extracts of star anise (*Illicium verum*) and some other plants like green anise (*Pimpinella anisum*) and fennel (*Foeniculum vulgare*). All these plants contain anethole as the main flavouring component. The final beverage is obtained by adding water to reduce the alcohol content till 35%~40% (v/v). The presence of metals in the final product comes from the raw materials, mainly plant seeds, ethyl alcohol and water or from the devices used in the elaboration process. Metals in alcoholic beverages are usually determined by atomic absorption or emission spectroscopic techniques. In many cases, a previous mineralisation step is necessary being this part of the procedure critical and very prone to introduce uncertainty into the final result. Anisette has organic matter coming from the plant extract. This organic matter has to be destroyed to avoid important matrix effects during the measurement step. Dry ashing (DA) or wet ashing (WA) with oxidizing reagents are the two most usual options.

Questions on Text

What is the relationship between Anisette and Marie Brizard?

Words to Watch

  matrix ['meɪtrɪks] n. 矩阵；模型；社会环境；基质

  aniseed ['ænɪsi:d] n. 茴香；八角

  pastis [pæ'sti:s] n. 法国茴香酒

  liquorice ['lɪkərɪs] n. 甘草；甘草糖

  dialect ['daɪəlɛkt] n. 方言，土话

  Salerno [sə'lɛrnoʊ] n. 萨莱诺

  ailment ['eɪlmənt] n. 轻病；小恙

  anethole [æ'nethəʊl] n. 茴香脑；对丙烯基茴香醚

  reagent [ri'eɪdʒənt] n. 试剂

Phrases and Patterns

  incline to 倾向于

  famous brand 名牌

  organic matter 有机物质

  raw material 原材料

Supplementary reading

  Marie Brizard was born on June 28, 1714 in Bordeaux, France. She is the third of 15 children of Pierre Brizard, barrel carpenter and his wife, Jeanne Laborde.

  On January 11, 1755, Marie found a West Indian sailor named Thomas, lying unconscious with high fever in the town square. Marie kindly took him home to take care of him and saved his life. In return, Thomas gave her the only treasure he possessed: the secret recipe of an extraordinary aniseed liqueur, which eventually led to create MARIE BRIZARD Anisette.

  With her nephew, Jean-Baptiste Roger, Marie founded Marie Brizard & Roger company the same year. The company began growing shortly after, launching other fine liqueurs including Parfait Amour, Fine Orange, Creme de Barbade, Cinnamon Liqueur, and Coffee Liqueur. Their products soon began gaining a quality reputation and were presented to Louis XV at Versailles by the Marshal of France, the Duke of Richelieu.

  By the death of Jean-Baptiste Roger in 1795, distillery was in full swing to meet ever growing demand. Six years later in 1801, Marie died at the age of 86. The company was then passed to Anne, the widow of Jean-Baptiste Roger, and then to her three children since Marie did not have any direct descendants (Jean-Baptiste Augustin, Basile Augustin and Theodore-Bernard Roger). The family kept its ownership of the company until 1998 over 10 generations and won numerous medals at prestigious national and international competitions and fairs for its superfine quality and reputation. In 1954, the company became a public limited company.

Today MARIE BRIZARD is available in over 120 countries with the reputation of highest quality liqueurs for the top bartenders. To support the global bartending community, the company has been hosting MARIE BRIZARD INTERNATIONAL BARTENDING SEMINAR AND COCKTAIL COMPETITION for over 25 years in Bordeaux, France, home of Marie Brizard. The brand continues to grow due to its strong presence among professional bartenders worldwide. Marie Brizard stays very close to its allies including the Association of Barmen in France (ABF), the International Barmen's Association (IBA), and the United States Bartenders' Guild (USBG).

Unit 2　Other Blended Alcoholic Beverages

Lesson 1　Vermouth

Objectives

At the end of this lesson, students should be able to:
1. Define Vermouth.
2. State two typical types of Vermouth.

Text

Vermouth is officially classified as an "aromatized fortified wine," referring to its derivation from a white base wine fortified and infused with a proprietary set of different plant parts: barks, seeds, and fruit peels. These are collectively termed botanicals. Vermouths are particularly popular in Europe and in the United States. The term "vermouth" is derived from the German word for wormwood Wermut. It is supposedly derived from Wer (man) and Mut (courage, spirit, manhood). When vermouth was introduced into Bavaria in the first half of the seventeenth century, by the Piedmont producer Alessio, *Artemisia absinthium* was probably translated literally as Wermutwein. When it reached France, it was changed to vermouth.

Vermouth is fortified up to 15%~21% alcohol. The proprietary mixture of herbs and spices impart an aromatic flavor as well as its bitter taste Vermouths are typically classified as sweet (Italian) or dry (French). In the Italian version, the alcohol content can vary from 15% to 17%, with 12%~15% sugar. French versions have 18% alcohol with 4% reducing sugar. Dry vermouth contains less herb and spice extract than the sweeter vermouth—about 3.74~5.62 mL/L for dry, and 5.62~7.49 mL/L for sweet.

Traditionally, vermouth and aperitif wines are prepared from grape-based wine, with the addition of an herb and spice mixture or their extracts. In Europe, these beverages are served straight (without the addition of water), whereas in America, they are mostly used in preparing cocktails. The herbal infusion gives vermouth its unique flavor and aroma.

The herbal infusions donate antioxidant characteristics to vermouths, primary from the addition of phenolic compounds. This may provide some protection against the oxidative stress.

A. Sweet (Italian) vermouth

Sweet vermouths are produced in Italy, Spain, and Argentina, as well as other countries, such as the United States. Typically, Italian vermouth is dark amber in color, with a light Muscat sweet nutty flavor. It also possesses a well-developed and pleasing fragrance, with a generous and warming taste, and a slightly bitter but agreeable aftertaste. In Italy, vermouth must contain at least 15.5% alcohol and 13% or more reducing sugar. American vermouths are generally higher in alcohol and somewhat lower in sugar.

B. Dry (French) vermouth

Dry vermouths usually have a higher alcohol content, lower sugar content, and are lighter color than sweet vermouths. In addition, they are usually more bitter in flavor. In a typical French dry vermouth, the alcohol content is 18% by volume, reducing sugar 4%, total acidity (as tartaric acid) 0.65%, and volatile acidity (as acetic acid) 0.053%.

Questions on Text

What is the main differences between Italian and French Vermouth?

Words to Watch

bark [bɑːk] n. 犬吠声；嗥叫声；树皮

wormwood [ˈwɜːmwʊd] n. 蒿，洋艾

manhood [ˈmænhʊd] n. 成年；成年时期；男儿气质

aperitif [əˈpɛrɪtɪf] n. 开胃酒

antioxidant [ˌæntɪˈɒksɪdənt] n. 抗氧化物质（如维生素 C 或 E，可消除体内自由基）

Phrases and Patterns

derive from 来源于

a set of 一套

vary from 不同于

by volume 按体积计算

tartaric acid 酒石酸

acetic acid 乙酸

Supplementary reading

Into the Twenty-first Century

Times were good for vermouth in France throughout the first half of the twentieth century. French vermouth was known worldwide, and often differentiated from Italian vermouth. In France, in 1930, vermouth de Chambéry had been granted protected status of origin designation, which was then called AO (Appellation d'Origine), a precursor to the French AOC (Appellation d'Origine Contrôllée) system of wine designations that came online in the 1930s. It appears that this designation had not been actively used for some time when, in 1990, European Union regulations required that all products submit applications to be reclassified under the EU'S (renamed) system, which in French is called AOP (Appellation d'Origine Protégée) and is sometimes rendered in English as PDO (Protected Designation of Origin).

But with the downturn in production in the late twentieth century, no companies submitted the paperwork for vermouth de Chambéry. So as of 1st July 2000, vermouth de Chambéry lost its protected designation status. If and when the Chambéry producers apply to reinstate the designation, it would still take a number of years for the application to wend its way through the EU organization, now that the original status has lapsed.

Because of vermouth's growing popularity, this application would seem a likely occurrence-but it very well may not happen at this point in time. Apparently, there must be more than one producer, and a certain amount of uniformity of production in a specified region, in order for a product to be granted protected designation of origin status. This status defines the geographic area, manufacturing methods and other key elements required for the product.

The two current Chambéry producers do not see eye-to-eye on the elements required for this application, despite the fact that the 'vermouth de Chambéry' designation could be even more significant when recent developments in Italy are considered. In March 2017 vermouth di Torino received its designated status in Italy, from the Italian government. With the accession to this status in Italy, the Italian designation is unlikely to be challenged in the EU. The first president elected was Roberto Bava of Cocchi and the vice president was Giorgio Castagnotti of Martini & Rossi. The founding members of the vermouth di Torino Institute are Berto, Bordiga, Del Professore, Carlo Alberto, Carpano, Chazalettes, Cinzano, Giulio Cocchi, Drapò, Gancia, La Canellese, Martini & Rossi, Sperone, Vergnano and Tosti.

This is by no means a complete list of Italian vermouth producers, or even of vermouth di Torino producers. There are new ones cropping up all the time, at distilleries and at wineries. These include Montanaro in Piedmont, made at a hundred-year-old grappa distillery near Alba that decided to expand their production range. And La Canellese, a production facility in the countryside outside of near the city of Asti, also serves as a production facility for vermouth brands from Italy and other countries.

The 'vermouth di Torino' (also spelled 'vermouth di Torino') designation celebrates the origins of vermouth in the city of Turin, and continues its formal status with the production rules by which the vermouth must be made. Specifically, it must be made in a designated area surrounding the cities of Turin and Asti, which are located in the Piedmont region of Italy.

Many small towns are included in this designated area, notably Canelli. This small city is located south of Asti and east of Alba—the two cities that also anchor Piedmont's famed wine production zone for the well-known red wines Barolo, Barbaresco, Dolcetto d'Alba and Ruché, and the white Roero and Moscato d'Asti wines. Today Canelli, with a population of only 10,000, is still a vibrant wine production-related industrial area, home to companies that produce labels, pallets, bottling machines and other supplies for winemakers who are headquartered nearby in Italy and in France.

One of the requirements of vermouth di Torino is that the grapes used to make the wine must be 75 percent local-grapes such as Moscato, also used to produce Piedmont's well-known sparkling Moscato d'Asti. The town of Canelli is thought to be a primary source of the historically significant Moscato grape that is known in English as Muscat Canelli. The name 'Muscat Canelli' is familiar to grape suppliers, though the grape itself may be more widely known to others as Muscat à Petits Grains; this grape has been planted in many countries for centuries so it is likely that just the English name, not the grape itself, originated in the town of Canelli.

Several large vermouth producers relocated their production facilities to the Canelli area when they grew too big to be headquartered in Turin in the nineteenth century. At that time, there was a busy railway line that came through Canelli's industrial zone. That rail service has lapsed and the former railway stations stand as charming, yet tantalizingly closed, lovely old buildings. However, several of the largest vermouth producers, such as Martini & Rossi and Gancia, are still located in their original warehouse spaces near the railway line.

Further north, on the way from Asti to Turin, the tiny town of Cocconato (population 1,600) is the current home of vermouth producers Cocchi and Les Chazalettes, as well as being the headquarters of the vermouth di Torino movement.

In Turin itself, there is currently only one vermouth producer, a new company established in 2012 called Drapò, a name that refers to the flag of Turin in the local dialect. This vermouth developed from the owner's hobby, with a family recipe from 1950 that grew so popular among his friends that he decided to create his own production facility. The feel at Drapò is super-artisanal. The base wines of Trebbiano and Moscato are locally sourced, as are many of the aromatic elements. Fruit, flower, herb and spice extractions are being fine-tuned with micro-extractions, pressure extractions and barrel-ageing of flavouring essences. The entire facility is postmodern in design: a Tuscan-gold, retro-hued exterior, large warehouse spaces with cheerful red floors and bright colours everywhere. There is even a small bar/classroom area for educational events and master classes.

Lesson 2 Martini

Objectives

At the end of this lesson, students should be able to:
1. Know what the main ingredients of Martini are.
2. Understand what the difference between Martini and Gibson is.

Text

Dry Martini: No cocktail recipe is more energetically argued over than the classic dry Martini. It is basically a generous measure of virtually neat stone-cold gin with a dash of dry white vermouth in it. But how much is a dash? Purists insist on no more than a single drop, or the residue left after briefly flushing the glass out with a splash of vermouth and then pouring it away. (They puzzlingly refer to such a Martini as "very dry", as if adding more vermouth would sweeten it. In fact, the terminology barks back to the original recipe, when the vermouth used was the sweet red variety.) Some go for as much as half a measure of vermouth, and I have at hand a book that suggests a two-to-one ratio of gin to vermouth-guaranteed to send the purist into paroxysms of horror. I have to admit I incline more to the purist philosophy, though: the vermouth should be added as if it were the last bottle in existence. The drink should properly be mixed gently in a separate jug, with ice, and then strained into the traditional cocktail glass (the real name of which is a martini glass). A twist of lemon peel should be squeezed delicately over the surface, so that the essential oil floats in globules on top of the drink, but don't put the lemon twist in the glass. And hold the olive. (Add a cocktail onion, however, and the drink becomes a Gibson.)

The American Standard dry martini shall come in the following three sizes: Regular: not less than 3.5 oz. Large: not less than 5 oz. Double: not less than 7 oz. Only the following three ingredients shall be used in the preparation of the American Standard dry martini: Gin, Dry vermouth and Olives. The employment of vermouth in an American Standard dry martini shall not be mandatory, provided no other ingredient is employed as a substitute, and the use of olives is not encouraged.

Questions on Text

What is the ratio of gin to vermouth is best for making Martini?

Words to Watch

purist [ˈpjʊərɪst] n. （语言等方面的）纯粹主义者
puzzlingly [ˈpʌzlɪŋli] adv. 莫名其妙地；使迷惑地
olive [ˈɒlɪv] n. 橄榄；橄榄树；橄榄色
energetically [ˌɛnəˈdʒɛtɪk(ə)li] adv. 精力充沛地；积极地
virtually [ˈvɜːtjʊəli] adv. 几乎；差不多；事实上
flush [flʌʃ] v. 脸红；冲洗
pour [pɔːr] v. 倾泻；浇注
terminology [ˌtɜːmɪˈnɒlədʒi] n. 术语；专门用语
paroxysm [ˈpærəksɪzm] n. 突然发作
jug [dʒʌg] n. 壶，罐
globule [ˈglɒbjuːl] n. 小滴，小球体

Phrases and Patterns

insist on 坚持
no more than 不过是；不超过
refer to 提到
in existence 存在
essential oil 精油
in accordance with 按照；符合

Supplementary reading

Early martinis were sweet enough to make your teeth ache (A maraschino cherry was used in the sweet martinis of yesteryear), but in 1896 came a recipe entailing two-thirds dry gin and one-third dry vermouth, which remained standard until the 1930s. Drier martinis, however, became fashionable, and dryness became a fetish. For instance, in *Across the River and Into the*

Trees (1949) Hemingway described the 15∶1 martini that he called the Montgomery after the favorable odds that the British general allegedly required before he would attack the enemy. As time passed, the main desideratum of the martini came to be dryness, and all sorts of methods were devised to keep the amount of vermouth to an absolute minimum. In the early 1950s the vermouth atomizer, which blew a mist of vermouth over the martini glass, came into use, and in the mid-1960s Hammacher Schlemmer introduced a long calibrated dropper designed to fit into a vermouth bottle, and syringes, scales (I received a scale as a gift from my mother-in-law and father-in-law; I used it once), and vermouth-infused stones appeared on the market. On August 16, 1963 Nicks Restaurant in Boston issued a commemorative placard claiming that its bartender had succeeded in isolating the vermouth molecule at noon that day.

In the 1957 play and 1967 musical Auntie Mame, based on Patrick Dennis's novel, young Patrick rinses the glass with a smidgeon of vermouth, which was then decanted, before adding iced gin, resulting in what was called the in-and-out martini. Another method was to pour the gin into an empty vermouth bottle and then over ice. Or, the vermouth was poured over ice cubes held in a sieve over a sink, and then the gin was poured over the same ice cubes into a pitcher. Or, the draft from an electric fan might be allowed to blow across the top of an open vermouth bottle in the direction of the pitcher or shaker. Or, the vermouth bottle might be placed next to the gin and the bottle turned slowly so that the label with the word vermouth was exposed to the gin for about a second. Or, the seeker of dryness might merely whisper the word vermouth over the gin or just salute in the general direction of France. The previously mentioned ASA document suggests: A 60watt incandescent lamp is placed on a flat surface 9 inches from a sealed bottle of vermouth. A sealed bottle of gin is placed on the other side of the bottle of vermouth at a distance of 23 inches. The lamp may be illuminated for an interval of 7 to 16 seconds. The duration of exposure is governed by the color of the bottles.

At the time of the first test of the nuclear bomb at the Trinity Site, Alamogordo Bombing Range, White Sands, New Mexico (5∶29 a.m., July 16, 1945), a bottle of vermouth was supposedly secreted in the plutonium bomb and subjected to nuclear fission. Thereafter one could merely hold the martini glass filled with gin out of the window to obtain a fissionable martini. Or, for the ultimate dry martini you might climb to Echo Point, shout Vermouth, and catch the echo in a glass of iced gin. Pure gin is sometimes called a naked martini.

Vocabulary

A

accelerate	v.	加速；加快
acetoin	n.	乙偶姻
acetolactate	n.	乙酰乳酸
acidity	n.	酸味；酸性
acidification power (AP)	n.	酸化力
acrolein	n.	丙烯醛
additive	n.	添加剂，食物添加剂；附加剂
adhumulone	n.	加葎草酮
adlupulone	n.	聚蛇麻酮；伴蛇麻酮
aeration	n.	通风，充气
ageing	n.	陈酿
airbag	n.	气囊
affordability	n.	支付能力；负担能力；可购性
afterthought	n.	随后添加的东西，追加物
agriculture	n.	农业；农学；农艺
ailment	n.	轻病；小恙
alcohol	n.	酒精，乙醇
alcoholic	adj.	酒精的；含酒精的；饮酒引起的
aldehyde	n.	醛类；乙醛
ale	n.	爱尔啤酒，艾尔啤酒
ambassadorial	adj.	大使的；使节的
amino acid	n.	氨基酸
amylase	n.	淀粉酶
anaerobic	adj.	厌氧的，厌气的
analysis of variance (ANOVA)	n.	方差分析
anethole	n.	茴香脑；对丙烯基茴香醚
aniseed	n.	茴芹籽

antioxidant	n.	抗氧化物质（如维生素 C 或维生素 E，可消除体内自由基）
antiradical power	n.	抗自由基能力
anthocyanogen	n.	花青素类
aperitif	n.	开胃酒
appreciation	n.	欣赏，鉴别；增值；感谢
approximately	adv.	大概；大约；约莫
aroma	n.	芳香
aromatic	adj.	芳香的，芬芳的；芳香族的
artificially	adv.	人工（造，为）地；不自然地
assessment	n.	看法；评估；评定
attribute	v.	把……归因于；
	n.	属性
authentic	adj.	真正的，真实的；可信的
authentication	n.	证明；鉴定；证实

B

bacillus	n.	杆菌，芽孢杆菌
bacteria	n.	细菌
bark	n.	树皮；吠声，嗥叫声；枪声
batch	n.	一批；一炉；一次所制之量
bentonite	n.	膨润土
berry	n.	浆果
beverage	n.	饮料
biomass	n.	生物质；生物量
biotechnological	adj.	生物技术的
blend	v.	混合；交融；掺和
boost	v.	推动；帮助；宣扬
Bordeaux	n.	波尔多（法国西南部港市）；波尔多葡萄酒
botanical	n.	植物学的
bottle	n.	（细颈）瓶子；一瓶（的量）
Burgundy	n.	勃艮第；勃艮第葡萄酒
butanol	n.	丁醇
2，3-butanediol	n.	2，3-丁二醇

by-product	n.	副产品;意外结果;副作用

C

Cabernet Sauvignon	n.	赤霞珠(红葡萄)
Cabernet Franc	n.	品丽珠;卡本内弗朗(红)
canonical variate analysis (CVA)	n.	规范变量分析
caramel	n.	焦糖;饴糖;焦糖糖果
carbohydrate	n.	碳水化合物;糖类
carbonyl	n.	羰基
carboxypeptidase	n.	羧肽酶
catechin	n.	儿茶酸;焦儿茶酸
cassava	n.	木薯
cereal	n.	谷类,谷物;谷类食品;谷类植物
champagne	n.	香槟酒
characteristic	adj.	典型的;特有的;表示特性的
Chardonnay	n.	霞多丽(白葡萄)
char	v.	炭化
cherry	n.	樱桃;樱桃树;如樱桃的鲜红色
chromatography	n.	色谱法,色层分析法,层析法
cinnamic	adj.	含苯乙烯基的
clarification	n.	澄清
climbing vine	n.	藤蔓
climate	n.	气候;气候区;倾向;思潮;风气;环境气氛
cocktail	n.	鸡尾酒
cohumulone	n.	合葎草酮
cognac	n.	科尼亚克白兰地酒,干邑
colupulone	n.	类蛇麻酮
combination	n.	结合;组合;联合;化合
component	n.	组成部分;成分
compound	n.	化合物;混合物
condensate	n.	冷凝物;浓缩物
condenser	n.	冷凝器;电容器
congener	n.	同种类或同性质的人或物
connoisseurship	n.	鉴赏能力

conscientious	adj.	勤勉认真的；一丝不苟的
consequent	adj.	随之发生的，作为结果的；（河流、山谷）顺向的；合乎逻辑的
consolation	n.	安慰；慰问；起安慰作用的人或事物
consumer	n.	消费者；用户，顾客
consumption	n.	消费；消耗
content	n.	内容，目录；满足；容量
conventional	adj.	依照惯例的；遵循习俗的；墨守成规的；传统的；习惯的
coordinate	v.	调节，配合；使动作协调
cordial	adj.	热情友好的；和蔼可亲的；
	n.	甜果汁饮料
corn	n.	（小麦等）谷物；谷粒
craftsmanship	n.	技术；技艺
critical	adj.	关键的；临界的；批评的；爱挑剔的；危险的；决定性的
corn grit	n.	玉米粉
crop	n.	庄稼；作物；（谷物、水果等一季的）收成，产量
crush	v.	破碎
cuisine	n.	烹饪；风味；饭菜，菜肴
cultivate	v.	耕作；种植；栽培；培育
custard	n.	蛋奶沙司；蛋奶糕，蛋挞

D

de-stemmed	adj.	除梗
deacidification	n.	脱酸
decarboxylase	n.	脱羧酶
decarboxylate	v.	脱羧基
decoction	n.	煮出法
degradation	n.	降解，退化；落泊，潦倒（的境况）；恶化（过程）
delicate	adj.	微妙的；精美的，雅致的；柔和的
descriptive analysis	n.	描述性分析；描述性统计

diacetyl	n.	双乙酰
dialect	n.	地方话；土话；方言
dictate	n. &v.	命令；指示
diastatic power	n.	糖化力
dioecious	adj.	雌雄异体的
diploid	n.	二倍体
distillation	n.	精馏，蒸馏，净化；蒸馏法；精华，蒸馏物
distinguished	adj.	卓越的，著名的；高贵的，受尊重的
distinguish	v.	区分；辨别；分清；成为……的特征
domestication	n.	驯化
dominant strain	n.	优势菌株
dosage	n.	（通常指药的）剂量
dry	v.	干燥
drunkenness	n.	醉态；酒醉
dry white wine	n.	干白葡萄酒

E

edible	adj.	可食用的
endogenous	adj.	内源性的
endosperm	n.	胚乳
energetically	adv.	精力充沛地；积极地
enhance	v.	提高；增强；增进
enzyme	n.	酶
equilibrium	n.	均衡
essence	n.	本质，实质；精华；香精
essential	adj.	基本的；必要的；本质的；精华的
esterification	n.	酯化（作用）
ethanol	n.	乙醇，酒精
ethanolysis	n.	乙醇分解
etiquette	n.	礼节，礼仪，规矩
evaluate	v.	估计；评价；评估
evaporation	n.	蒸发

F

fermentation	n.	发酵

fiber	n.	纤维
fiber dietary	n.	膳食纤维
filter	v.	过滤，滤除
fingerprint	n.	指纹；指印
flavor	n.	风味；香料；滋味
flush	v.	发红；脸红；冲洗
flocculation	n.	絮凝；凝聚
foreshot	n.	（蒸馏威士忌酒时）初馏分，酒头
formation	n.	组成；形成；组成物；形成物；编队；队形
fraction	n.	分数；小部分；稍微
fragrance	n.	香味，芬芳；香料；香精
friability	n.	易碎性
furan	n.	呋喃；氧杂茂
furfural	n.	糠醛；呋喃甲醛

G

gas chromatography（GC）	n.	气相色谱分析
geographical	adj.	地理的；地理学的
geographically	adv.	地理学上
germinate	v.	发芽，萌芽，开始生长
germination	n.	发芽
globule	n.	小滴，小球体
glucoamylase	n.	糖化酶；葡萄糖淀粉酶
glycerin	n.	甘油
grown	adj.	成熟的；成年的；长大的

H

harmful	adj.	（尤指对健康或环境）有害的，导致损害的
harmonious	adj.	和谐的，和睦的；协调的；悦耳的
herb	n.	药草；香草；草本植物
hexanoate	n.	己酸盐
hexaploid	n.	六倍体
hop cone	n.	蛇麻球果；酒花果穗
hop resin	n.	酒花树脂

horizontal	adj.	水平的；与地面平行的；横的
humulone	n.	葎草酮
Humulus	n.	葎草属
Humulus lupulus L.	n.	啤酒花（大麻科葎草属）
hydrolyse	v.	水解
hydrolase	n.	水解酶
hyperalgesia	n.	痛觉过敏

I

incubation	n.	孵化
infusion	n.	注入；灌输
ingredient	n.	原料；要素；组成部分
inherent	adj.	固有的；内在的；与生俱来的，遗传的
inoculation	n.	接种
insufficient	adj.	不充分的；不足的；不够重要的
intellectual	adj.	智力的；聪明的；理智的
inventory	n.	存货，存货清单；详细目录；财产清册
isoamyl	n.	异戊基

J

jiuqu	n.	酒曲
jug	n.	壶，罐
juniper	n.	杜松；桧或刺柏属植物
jurisdiction	n.	司法权，审判权，管辖权

K

ketone	n.	酮类
koji	n.	曲；日本酒曲；清酒曲

L

lactate	n.	乳酸盐
lactic acid bacteria（LAB）	n.	乳酸菌
lactobacillus	n.	乳杆菌
lactone	n.	内酯

lager	n.	拉格啤酒,贮藏啤酒
lees	n.	酒脚,酒泥,酒底沉淀物
lengthwise	adv.	纵向地;纵长地
leveraging	n.	杠杆作用;杠杆
libation	n.	饮酒;祭酒
lignin	n.	木质素
lipid	n.	脂肪,油脂
lipoxygenase activity	n.	脂氧合酶活性
liqueur	n.	利口酒,香甜酒
liquor	n.	酒,含酒精饮料;烈酒
liquorice	n.	甘草;甘草糖
lupulone	n.	蛇麻酮

M

macromolecular	adj.	大分子的
Maillard reaction	n.	美拉德反应
malolactic fermentation	n.	苹果酸-乳酸发酵
manganese	n.	锰
mangle	v.	压碎;撕烂;严重损坏;糟蹋
manhood	n.	成年;成年时期;男儿气质
mash	n.	糖化醪
matrix	n.	矩阵;模型;社会环境;基质
melon	n.	甜瓜;瓜
metabolic	adj.	新陈代谢的;代谢的
metabolite	n.	代谢物;代谢分子
methanol	n.	甲醇
methional	n.	甲硫基丙醛
Merlot	n.	美乐(红葡萄)
microbiota	n.	小型生物群,微生物区
microorganism	n.	微生物
millet	n.	粟;谷子;黍类
mineral	n.	矿物质
mix	v.	(使)混合;配制;参与
molasses	n.	糖蜜,糖浆

mold	n.	霉菌
molecule	n.	分子，微粒
moreover	adv.	而且；此外
mould	n.	霉，霉菌
must	n.	待发酵葡萄汁

N

neutral	adj.	中立的，中性的；中立国的；非彩色的
niacin	n.	烟酸
nonenal potential	n.	壬烯醛潜力
nutmeg	n.	肉豆蔻
nutrition	n.	营养，营养学；营养品
nutritional	adj.	营养的；滋养的

O

oaky aromas	n.	橡木桶香气
oenococcus	n.	酒类酒球菌
oenological	adj.	有关酒类研究的
olive	n.	橄榄；橄榄树
optimal	adj.	最优的；最佳的
optimization	n.	最佳化，最优化
organic	adj.	有机的；不使用化肥的；绿色的；有机物的；生物的；器官的；器质性的；官能的
organic acid	n.	有机酸
organism	n.	有机体；生物；有机组织；有机体系
organoleptic	adj.	感官的；用感官觉察的
original	n.	原件；原作；原物；原型
origin	n.	起源
osmotic	adj.	渗透性的

P

package	n.	包装
paradoxically	adv.	自相矛盾地；似是而非地；反常地
parameter	n.	参数，参量，变量

paroxysm	n.	突然发作
pasteurize	v.	用巴氏杀菌法给……消毒
pastis	n.	法国茴香酒
peptide	n.	肽；缩氨酸
percolation	n.	过滤；浸透
pectinase	n.	果胶酶
perennial	adj.	长久的；持续的；反复出现的；多年生的
perlite	n.	珍珠岩
perspective	n.	观点；远景；透视图
phenolic	adj.	酚的
Phosphorus	n.	磷
Physicochemical	adj.	物理化学的
pigment	n.	色素
Pinot Noir	n.	黑比诺（红葡萄）
pitch	n.	运动场地；程度；音高
	v.	给（麦芽汁）加酵母
plum	n.	李子；梅子
Poaceae/Gramineae	n.	禾本科
polyphenol	n.	多酚
polysaccharide	n.	多糖
portfolio	n.	公文包；文件夹；（公司或机构提供的）系列产品，系列服务
potency	n.	效能；力量；潜力；权势
pour	n.	倾倒；倒出
precursor	n.	先驱；先锋；前身
press	v.	压榨
pretreatment	n.	预处理；预处理期间的
primary fermentation	n.	初次发酵，主发酵
principal component analysis (PCA)	n.	主成分分析
production	n.	生产；制造；制作；产量；产生，分泌
proof	n.	证明；证据
proportion	n.	比例，占比；部分；面积；均衡
protease	n.	蛋白酶
protein	n.	蛋白质

pulp	n.	果肉；黏浆状物质
pungent	adj.	辛辣的；刺激性的；刺鼻的；苦痛的；尖刻的
purification	n.	净化；提纯
purist	n.	（语言等方面的）纯粹主义者
puzzlingly	adv.	莫名其妙地；使迷惑地

R

reagent	n.	试剂
recipe	n.	配方，处方
rectification	n.	精馏
reductase	n.	还原酶
reducing power	n.	还原能力
refrigerate	v.	使冷却；使变冷；冷藏
relative density	n.	相对密度
relatively	adv.	相当程度上；相当地；相对地
removal	n.	移动；去除
represent	v.	代表；作为……的代言人；维护……的利益；等于；相当于；意味着
requirement	n.	要求；必要条件；必需品
Riesling	n.	雷司令（白葡萄）
rival	n.	竞争对手；可与……匹敌的人；同行者
rye	n.	黑麦

S

saccharose	n.	蔗糖
saccharification	n.	糖化（作用）
Saccharomyces cerevisiae	n.	酿酒酵母
Sauvignon Blanc	n.	长相思（白）
scope	n.	范围；余地；视野；眼界；导弹射程
scrape	v.	刮掉；削去
seasoning	n.	调味品；佐料
secondary metabolite	n.	次级代谢产物
sediment	n.	沉淀物；沉积物

Semillon	n.	赛美蓉（白）
sensory	adj.	感觉的；感官的
separation	n.	分离；分开；分割
simultaneous	adj.	同时的；联立的；同时发生的
smoothness	n.	平滑；柔滑；平坦
softer	adj.	柔和的
sorbitol	n.	山梨醇
sorghum	n.	高粱；蜀黍
sorption	n.	吸附作用；吸着作用；吸收作用
source	n.	来源
sparkling wine	n.	起泡酒
spectrometric	adj.	光谱测定的；分光仪的；能谱仪的
specialty beer	n.	特色啤酒
spiced	adj.	五香的；调过味的；含香料的
spoilage	v.	有害；（食物的）变质，腐败
spontaneous fermentation	n.	自然发酵
starch	n.	淀粉；含淀粉的食物
stability	n.	稳定性
stalk removal	n.	除茎
storage	n.	存储；仓库；贮藏所
strain	n.	菌株
strip	n.	带，条状物
structure	n.	结构；构造
substance	n.	物质；实质
substrate	n.	底物；底层；基底；基层
sugar	n.	食糖；一匙糖；一块方糖；（植物、水果等所含的）糖
sulfite	n.	亚硫酸盐
sweetness	n.	令人愉快；讨人喜欢；甜；芬芳

T

tannin	n.	单宁；鞣质
temperature	n.	温度；体温；气温
terminology	n.	术语；有特别含义的用语；专门用语

tetraploid	n.	四倍体
threshold	n.	阈值；界；起始点
traditional	adj.	传统的；惯例的
tropical fruit	n.	热带水果
trub	n.	凝固物，冷却残渣

U

ultrasonic	adj.	超声的
undergo	v.	经历，经受
undrinkable	adj.	不能喝的；不能饮用的
unmistakable	adj.	明显的；不会弄错的
Urticaceae	n.	荨麻科

V

value	n.	（数学中的）值，（商品的）价值
vanilla	n.	香草香精
vanillin	n.	香草醛
vapour	n.	蒸气；潮气；雾气
variability	n.	可变性；易变性；反复不定
versatility	n.	多功能性；多才多艺；用途广泛
vibration screening	n.	振动筛选
virtually	adv.	几乎；差不多；事实上
vitamin	n.	维生素
viticulture	n.	葡萄栽培学
viticultural	adj.	葡萄栽培的
volatile	adj.	易挥发的；易变的；不稳定的
volatile organic compounds（VOC）	n.	挥发性有机物
volatility	n.	挥发性；易变
volatilization	n.	蒸发，挥发；发散
volumetric	adj.	体积的；容积的；测定体积的
vulnerable	adj.	（身体上或感情上）脆弱的，易受伤害的

W

whirlpool	n.	回旋槽；旋涡

wormwood	n.	苦蒿，苦艾

X

xylose	n.	木糖

Y

yeast	n.	酵母
yeast starter	n.	酵母发酵剂
yolk	n.	蛋黄

References

[1] Almaguer C, Schönberger C, Gastl M, et al. *Humulus lupulus*-a story that begs to be told. A review [J]. Journal of the institute of brewing, 2014, 120 (4): 289-314.

[2] Alves V, Gonçalves J, Figueira J A, et al. Beer volatile fingerprinting at different brewing steps [J]. Food chemistry, 2020: 326.

[3] Appel J M. "Physicians are not bootleggers": the short, peculiar life of the medicinal alcohol movement [J]. Bulletin of the History of Medicine, 2008, 82 (2): 355-86.

[4] Brick J. The World's Best Rum Comes From These Countries. 2016. https://www.thrillist.com/drink/nation/best-rum-regions-worlds-best-rum.

[5] Bamforth C W. Brewing New technologies. London: Woodhead Publishing Limited, 2006.

[6] Cai H, Zhang T, Zhang Q, et al. Microbial diversity and chemical analysis of the starters used in traditional Chinese sweet rice wine. Food Microbiology, 2018, 73: 319-326.

[7] Cao Y, Xie G, Wu C, et al. A Study on Characteristic Flavor Compounds in Traditional Chinese Rice Wine-Guyue Longshan Rice Wine [J]. Journal of the Institute of Brewing, 2010, 116 (2): 182-189.

[8] Chen A J, Fu Y Y, Jiang C, et al. Effect of mixed fermentation (Jiuqu and *Saccharomyces cerevisiae* EC1118) on the quality improvement of kiwi wine. CyTA-Journal of Food, 2019, 17 (1): 967-975.

[9] Comuzzo P, Rauhut D, Werner M, et al. A survey on wines from organic viticulture from different European countries [J]. Food Control, 2013, 34 (2): 274-282.

[10] Daniel O, Perez-Correa JR, Lorenz T B, et al. Wine Distillates: Practical Operating Recipe Formulation for Stills. Journal of Agricultural Food Chemistry. 2005, 53: 6326-6331.

[11] Liu D F, Zhang H T, Xiong W L, et al. Effect of Temperature on Chinese Rice Wine Brewing with High Concentration Presteamed Whole Sticky Rice [J]. Biomed Research International, 2014, 2014: 1-8.

[12] Duc-Truc P, Renata R, Vanessa J S, et al. Influence of partial dealcoholization on the composition and sensory properties of Cabernet Sauvignon wines. Food Chemistry, 2020, 325 (Sep. 30): 126869.1-126869.8.

[13] Duncan, N A, Starbuck J, Liu L. A Method to Identify Cross-Shaped Phytoliths of Job's Tears, Coix lacryma-jobi L., in Northern China. J. Archaeolog. 2019, 24: 16-23.

[14] Ekelöf E. Medical aspects of the Swedish Antarctic Expedition October 1901-January 1904. J Hyg. 1904, 4: 511-40.

[15] Epstein B S. Strong, Sweet & Dry: A GUIDE TO VERMOUTH, PORT, SHERRY, MADEIRA AND MARSALA. London: Reaktion Books Ltd, 2020.

[16] Fernández-Pérez R, Rodríguez C T, Fernanda R L. Fluorescence microscopy to monitor wine malo-

lactic fermentation [J]. Food Chemistry, 2019, 274 (Feb. 15): 228-233.

[17] Foss R. Rum: A Global History. London: Reaktion Books, 2012.

[18] Gallone B, Steensels J, Prahl T, et al. Domestication and Divergence of Saccharomyces cerevisiae Beer Yeasts [J]. Cell, 2016, 166 (6): 1397-1410.

[19] Gaytán M S. Tequila! Distilling the Spirit of Mexico. California: Stanford University Press, 2014.

[20] Gianvito P D, Perpetuini G, Tittarelli F, et al. Impact of Saccharomyces cerevisiae strains on traditional sparkling wines production [J]. Food research international, 2018, 109 (Jul.): 552-560.

[21] Godtfredsen S E, Ottesen M. Maturation of beer with α-acetolactate decarboxylase [J]. Carlsberg Research Communications, 1982, 47 (2): 93-102.

[22] Guido, Luís F, Curto A F, Boivin P, et al. Correlation of Malt Quality Parameters and Beer Flavor Stability: Multivariate Analysis [J]. Journal of Agricultural and Food Chemistry, 2007, 55 (3): 728-733.

[23] Hannemann, Wolfgang. Reducing Beer Maturation Time and Retaining Quality [J]. Technical Quarterly, 2002.

[24] Hein K, Ebeler S E, Heymann H. Perception of fruity and vegetative aromas in red wine [J]. Journal of Sensory Studies, 2009, 24 (3): 441-455.

[25] Hisashi Y, Masafumi T, Hajime K, et al. Investigation of relationship between sake-making parameters and sake metabolites using a newly developed sake metabolome analysis method. Journal of Bioscience and Bioengineering, 2019, 128: 183-190.

[26] Hisashi Y, Masafumi T, Hajime K, et al. Investigation of relationship between sake-making parameters and sake metabolites using a newly developed sake metabolome analysis method. Journal of Bioscience and Bioengineering, 2019, 128: 183-190.

[27] Ilda C, Rui S, Jorge M R. Kinetics of odorant compounds in wine brandies aged in different systems. Food Chemistry. 2016: 15, 46-48.

[28] Ivit N N, Loira I, Morata A, et al. Making natural sparkling wines with non-Saccharomyces yeasts [J]. European Food Research & Technology, 2018, 244 (5): 925-935.

[29] Jeng J S, Siao H Y. Acceleration of the Aging Process in Coffee Liqueur by Ultrasonic Wave Treatment. 2016, 40 (3): 502-508.

[30] Jin G Y, Zhu Y, Xu Y. Mystery behind Chinese liquor fermentation. Trends in Food Science & Technology, 2017, 63: 18-28.

[31] Jurado J M, Alcázar A, Pablos F, et al. Determination of Zn, B, Fe, Mg, Ca, Na and Si in anisette samples by inductively coupled plasma atomic emission spectrometry. Talanta. 2004. doi: 10.1016/j.talanta.2003.10.052.

[32] Kaneda H, Kano Y, Sekine T, et al. Effect of pitching yeast and wort preparation on flavor stability of beer [J]. Journal of Fermentation & Bioengineering, 1992, 73 (6): 456-460.

[33] Kato Y. History of Marie Brizard. 2018. Retrieved from http://www.cocktailtimes.com/mariebrizard/press/mariebrizard_history.pdf.

[34] Kauffman G B. The Dry Martini: Chemistry, History, and Assorted Lore. The Chemical Educator.

2001, 6, 295-305. doi: 10. 1007/s00897010505a.

[35] Lehnhardt F, Gastl M, Becker T. Forced into aging: Analytical prediction of the flavor-stability of lager beer. A review [J]. Critical Reviews in Food Science & Nutrition, 2018: 1-35.

[36] Lehnhardt F, Becker T, Gastl M. Flavor stability assessment of lager beer: what we can learn by comparing established methods. European food research and technology, 2020, 246: 1105-1118.

[37] Liu S P, Mao J, Liu Y Y. Bacterial succession and the dynamics of volatile compounds during the fermentation of Chinese rice wine from Shaoxing region. World J Microbiol Biotechnol, 2015, 31: 1907-1921.

[38] Madrera R R, Gomis D B, Alonso J J. Influence of distillation system, oak wood type, and aging time on volatile compounds of cider brandy. Journal of Agricultural and Food Chemistry. 2003, 51 (19): 5709-5714.

[39] Miller G H. Whisky Science. Berlin: Springer International Publishing, 2019.

[40] Misaki A, Hiroto M, Yukiko K, et al. Chemical and Bacterial Components in Sake and Sake Production Process. Current Microbiology, 2019, 116: 56-59.

[41] Monica S M. Carmen R, Manuel S, et al. Development of an accelerated aging method for Brandy. LWT-Food Science and Technology. 2014, 59: 108-114.

[42] Morris R. The Joy of Home Distilling: The Ultimate Guide to Making Your Own Vodka, Whiskey, Rum, Brandy, Moonshine, and More. New York: Skyhorse Publishing, 2014.

[43] Nana M, Yoshiro T, Samanthi W P. Effects of Sake lees (Sake-kasu) supplementation on the quality characteristics of fermented dry sausages. Heliyon, 2020, 6: 15-18.

[44] Naseer A, Harmeet C, Babita A, et al. Review on Development of Wine and Vermouth from the Blends of Different Fruits. Journal of Food Processing & Technology, 2017, 8 (1): 1-7.

[45] Panesar P S, Joshi V K, Panesar R, et al. Vermouth: Technology of Production and Quality Characteristics. Advances in Food and Nutrition Research. Elsevier Inc, 2011.

[46] Pascali S A D, Gueletta A, Coco L D, et al. Viticultural practice and winemaking effects on metabolic profile of Negroamaro [J]. Food Chemistry, 2014, 161 (Oct. 15): 112-119.

[47] Paterson A, Swanston J S, Piggott J R, et al. Fermented Beverage Production (2nd ed.). New York: Springer US, 2003.

[48] Piggott J R. Whisky, Whiskey and Bourbon: Composition and Analysis of Whisky. Encyclopedia of food and health. 2016, doi: 10. 1016/b978-0-12-384947-2. 00752-2.

[49] Qian M C, Shellhammer T H. Flavor Chemistry of Wine and Other Alcoholic Beverages. American Chemical Society, 2012.

[50] Roald A. Roald Amundsen's Belgica diary. Huntingdon: Bluntisham Books, 1999.

[51] Rogns G H, Rathe M, Petersen M A, et al. From wine to wine reduction: Sensory and chemical aspects [J]. International Journal of Gastronomy and Food Science, 2017, 9: 62-74.

[52] Russell I, Stewart G. Whisky: Technology, Production and Marketing. London: Elsevier Academic Press, 2015.

[53] Śliwińska M, Wiśniewska P, Dymerski T. The flavour of fruit spirits and fruit liqueurs: a review. Flavour and Fragrance Journal. 2014, doi: 10.1002/ffj.3237.

[54] Shyr J J, Yang S H. Acceleration of the aging process in coffee liqueur by ultrasonic wave treatment. Journal of Food Processing and preservation, 2015.

[55] Sigler K, Matoulková D, Dienstbier M, et al. Net effect of wort osmotic pressure on fermentation course, yeast vitality, beer flavor, and haze [J]. Applied Microbiology & Biotechnology, 2009, 82 (6): 1027-1035.

[56] Solmonson L J. Gin: A Global History. London: Reaktion Books, 2012.

[57] Sumby K M, Bartle L, Grbin P R, et al. Measures to improve wine malolactic fermentation [J]. Applied Microbiology & Biotechnology, 2019, 103 (5): 2033-2051.

[58] Toh D W K, Chua J Y, Lu Y, et al. Evaluation of the potential of commercial non - *Saccharomyces* yeast strains of *Torulaspora delbrueckii* and *Lachancea thermotolerans* in beer fermentation [J]. International Journal of Food Science & Technology, 2019.

[59] Walton S, Glover B. The ultimate encyclopedia of wine, beer, spirits & liqueurs. London: Hermes House, 2014.

[60] Wei X L, Liu S P, Yu J S, et al. Innovation Jimo Laojiu rice wine brewing technology by bi-acidification to exclude ice soaking process. J Biosci. Bioeng. 2017, 123: 460-465.

[61] Williams I. Tequila: A Global History. London: Reaktion Books, 2015.

[62] Yang Y, Xia Y, Wang G, et al. Effect of mixed yeast starter on volatile flavor compounds in Chinese rice wine during different brewing stages [J]. Lwt Food Science & Technology, 2017, 78: 373-381.

[63] Yu Q Q, Zhang Y C, Jiao Z G, et al. Key techniques and process optimization of blending brandy. Sino-overseas grapevine & wine. 2018, 6: 13-15.

[64] Yue Y Y, Zhang W X, Yang R, et al. Design and Operation of an Artificial Pit for the Fermentation of Chinese Liquor. Journal of the institute of brewing, 2007, 113: 374-380.

[65] Zhang B, Guan Z B, Cao Y, et al. Secretome of *Aspergillus oryzae* in Shaoxing rice wine koji. International Journal of Food Microbiology, 2012, 155: 113-119.

[66] Zhang B, Zeng X A, Sun D W, et al. Effect of Electric Field Treatments on Brandy Aging in Oak Barrels. Food and Bioprocess Technology, 2013, 6: 1635-1643.

[67] Zhang B, Kong L Q, Cao Y, et al. Metaproteomic haracterisation of a Shaoxing rice wine "wheat Qu" extract. Food Chemistry, 2012, 134: 387-391.

[68] Zhang C L, Ao Z H, Chui W Q, et al. Characterization of volatile compounds from Daqu-a traditional Chinese liquor fermentation starter. International Journal of Food Science and Technology, 2011, 46: 1591-1599.

[69] Zhang X, Yu Z, Tang W, et al. Improvement of barley (*Hordeum vulgare* L.) germination by application of biochar leacheate in steeping solution to upgrade malt quality [J]. Biotechnology Letters, 2020, 42 (2): 305-311.

[70] Zhong J J, Ye XQ, Fang Z X. Determination of biogenic amines in semi-dry and semi-sweet Chi-

nese rice wines from the Shaoxing region. Food Control, 2012, 28: 151-156.

[71] Zhu Y B., Zhang J H., Shi Z P, et al.: Optimization of operating conditions in rice heat blast process for Jimo Laojiu rice wine production by combinational utilization of neural network and genetic algorithms. J. Inst. Brew. 2004, 110: 117-123.